Nerve Cells and Animal Behaviour
Second Edition

This new edition of **Nerve Cells and Animal Behaviour** has been updated and expanded by Peter Simmons and David Young in order to offer a comprehensive introduction to the field of neuroethology while still maintaining the accessibility of the book to university students. Two new chapters have been added, broadening the scope of the book by describing changes in behaviour and how networks of nerve cells control behaviour.

The book explains the way in which the nervous systems of animals control behaviour without assuming that the reader has any prior knowledge of neurophysiology. Using a carefully selected series of behaviour patterns, students are taken from an elementary-level introduction to a point at which sufficient detail has been assimilated to allow a satisfying insight into current research on how nervous systems control and generate behaviour. Only examples for which it has been possible to establish a clear link between the activity of particular nerve cells and a pattern of behaviour have been used.

Important and possibly unfamiliar terminology is defined directly or by context when it first appears and is printed in bold type. At the end of each chapter, the authors have added a list of suggestions for further reading, and specific topics are highlighted in boxes within the text.

Nerve Cells and Animal Behaviour is essential reading for undergraduate and graduate students of zoology, psychology and physiology and serves as a clear introduction to the field of neuroethology.

PETER SIMMONS is a Lecturer in the Department of Neurobiology, University of Newcastle upon Tyne, UK, and DAVID YOUNG is a Reader in the Department of Zoology, University of Melbourne, Australia. Both authors regularly publish their research in insect neuroethology.

Nerve cells and animal behaviour

SECOND EDITION

PETER J SIMMONS and DAVID YOUNG

CAMBRIDGE
UNIVERSITY PRESS

PUBLISHED BY THE PRESS SYNDICATE OF THE UNIVERSITY OF CAMBRIDGE
The Pitt Building, Trumpington Street, Cambridge, United Kingdom

CAMBRIDGE UNIVERSITY PRESS
The Edinburgh Building, Cambridge CB2 2RU, UK   www.cup.cam.ac.uk
40 West 20th Street, New York, NY 10011-4211, USA   www.cup.org
10 Stamford Road, Oakleigh, Melbourne 3166, Australia
Ruiz de Alarcón 13, 28014 Madrid, Spain

First edition published 1989
Second edition published 1999

Printed in the United Kingdom at the University Press, Cambridge

Typeface Utopia 9.25/14 pt.   *System* QuarkXPress®   [SE]

A catalogue record for this book is available from the British Library

Library of Congress Cataloguing in Publication data

Simmons, Peter (Peter John), 1952–
Nerve cells and animal behaviour / Peter Simmons and David Young.
– 2nd ed.
    p.   cm.
Rev. ed. of: Nerve cells and animal behaviour / David Young. 1989.
ISBN 0 521 62216 6 (hardcover)
1. Neurobiology.   2. Neurons.   3. Animal behavior.
4. Neurophysiology.   I. Young, David, 1942 Feb. 25–   II. Young,
David, 1942 Feb. 25–.   Nerve cells and animal behaviour.
III. Title.
QP356.Y68   1999
573.8′6–dc21   99-11620 CIP

ISBN 0 521 62216 6 hardback
ISBN 0 521 62726 5 paperback

CONTENTS

PREFACE

Our aim in this book is to introduce university students to research on nervous systems that is directly relevant to animal behaviour, and to do so at a level that assumes no detailed knowledge of neurophysiology. Many topics that fall within the scope of neurobiology are omitted or passed over lightly, and attention is concentrated on particular examples that illustrate clearly how the activity of nerve cells is linked with animal behaviour. Since the first edition was published, many new books on neurobiology have appeared, but most concentrate on the cellular and physiological aspects of the nervous system. By reviewing some of the modern stories in neuroethology, we hope that this book will also be useful to postgraduate students and others who wish to learn something of the way in which behaviour is controlled.

Each major topic in Chapters 3–9 is dealt with as far as possible by introducing a particular type of behaviour and then working towards a description of how nerve cells control it. We have selected subjects from studies in which the links between nerve cells and animal behaviour are particularly clear. In doing this, we hope to illustrate the principles that have been revealed in modern research in neuroethology. Inevitably, there are many interesting stories that we have not been able to touch upon.

Readers who are familiar with the first edition of the book will notice several changes in content and arrangement. The final two chapters, on circuits of nerve cells and on plasticity in behaviour, are completely new. In order to provide an early illustration of how activity in nerve cells can be related to animal behaviour, we now describe work on prey detection by toads in the first chapter, and the chapter on startle behaviour is placed earlier in the book than it was in the previous edition. New material has been added in several places, particularly in Chapters 3, 5 and 7. In order to make room for this new material, we have had to omit the chapter on intraspecific communication.

To help readers to come to grips with unfamiliar terms and concepts, many of these are set in **bold type** the first time they appear in the book, as well as in the index. Anyone who studies neurobiology will soon discover that it has many side branches, linking one story to another or to other branches of biology. We have included brief introductions to a few of these by means of boxes in some chapters. These boxes do not have to be read as part of the main text, but are meant to complement it by providing useful and interesting, relevant information.

Suggestions for further reading are given at the end of each chapter, and major references to points of detail are scattered through the text and listed at the end of the book. The references in the figure legends also draw attention to relevant papers as well as indicating our grateful acknowledgement of material from other authors that we have incorporated into the figures.

We would like to thank many colleagues who have given useful comments on various aspects of the book, particularly Claire Rind and a number of undergraduate and postgraduate students. We are also very grateful to members of our families for their support during the preparation of the book.

1 Introduction

1.1 Nervous systems and the study of behaviour

People in antiquity seem to have had no idea that the brain was in any way connected with behaviour. Even that great practical biologist Aristotle was mistaken in his ideas. He observed the rich vascular supply of the brain and concluded that it was an organ for cooling the blood. The ancient Egyptians were positively cavalier in their attitude: when the body of a monarch was being prepared for mummification, the brain was extracted with a spoon and thrown away. The brain was considered unnecessary for the future life, but the entrails were carefully preserved in a jar and kept beside the mummified body.

Modern opinion emphasises the paramount importance of the brain as the source of an individual's behaviour and personality. This trend has gone so far that many a successful work of science fiction has been based on the idea that the brain might be kept alive or transplanted, and that by this means the essential personality of the original individual might be preserved after the rest of the body has been disposed of. This vast change in prevailing opinion about the brain is, of course, due to the anatomical and physiological research of the last 200 years, which has revealed the nature and importance of the central nervous system.

Our present understanding of the way in which nervous systems control animal behaviour owes much to a group of biologists working in the middle of the twentieth century, who pioneered an experimental approach to analysing behaviour. The approach they adopted came to be known as **ethology**, and one of early ethology's most thoughtful exponents was Niko Tinbergen. In an important paper, 'On aims and methods of ethology' (Tinbergen, 1963), he defined ethology simply as 'the biological study of behaviour'.

Tinbergen himself made an impact on ethology by concentrating on field observations or on elegantly simple experiments carried out on intact animals. But he expected that the results of this work would be integrated with a neural analysis as this became available. This is seen clearly in his book synthesising ethology, entitled *The Study of Instinct* (1951), in which he referred to contemporary research in neurophysiology and formulated his concepts in terms of the nervous system as far as possible. He expected that the biological methods of ethology would yield 'concrete problems that can be tackled both by the ethologist and the physiologist', and he wrote of 'the fundamental identity of the neurophysiological and the ethological approach'.

The long-term goal of such an approach is to analyse patterns of behaviour in terms of the activity of the underlying neural components. Hence, this field of research is sometimes given the title of **neuroethology**, a term that first came into use in the 1960s. Neuroethology tries to combine the approaches of both ethology and neurobiology so as to understand the neural basis of behaviour. Often, this involves examining groups of receptors or networks of nerve cells in order to elucidate the interactions relevant to behaviour. In some cases it is possible to bring both neurobiological and ethological analysis to bear on a single phenomenon, as Tinbergen expected.

In the chapters that follow, selected examples are considered in which neural analysis has been carried out in a way that is helpful to an understanding of animals' natural behaviour. As far as possible, attention is concentrated on specific case histories in which a connection has been established between a particular group of nerve cells (also termed neurons) and a particular pattern of behaviour. This field of study is developing rapidly and enough has been accomplished to enable initial conclusions to be drawn about the operation of many basic areas. These studies and conclusions form an essential and fascinating part of ethology, the biological study of behaviour.

1.2 Scope and limitations of neuroethology

As might be expected, neuroethology has been most successful in tackling those elementary components of behaviour with which ethology itself began. The simple kinds of behaviour that first caught the attention of the founders of ethology are often also the kinds of behaviour most readily analysed in terms of the underlying neural events. One good example is intraspecific communication, which requires both that the sender delivers a clear signal and that the receiver has the appropriate sensory apparatus to analyse it. The interactions between predator and prey have also been the subject of many neuroethological studies because the neural mechanisms involved must be simple in order to be swift. When life-or-death decisions have to be made in a small fraction of a second, there is just not time for elaborate neural circuits to operate.

A good many of the cases that have been analysed successfully involve **dedicated systems**. A dedicated neural system is one that is largely devoted to a single, important function such as escape (see Chapter 3). Dedicated systems are easier to analyse than multipurpose systems, not merely because they tend to be simpler, but more importantly because their behavioural function is clearly known. As neural systems become more flexible in the tasks that they can perform, it becomes more difficult for experimenters to determine what is behaviourally important in their neurophysiological recordings. In a multipurpose system, it is difficult to discern which of several possible functions is pertinent to neural activity recorded in a dissected animal. In a dedicated system, any recorded activity is likely to relate to the one and only behavioural function, provided the system is in a healthy state.

For example, the large amount of neurophysiological work that has been done on hearing in cats has been of little interest to ethologists because it is so difficult to correlate particular properties of the auditory system with particular episodes in the animal's normal behaviour. It is almost impossible to know what a cat is listening to at any given moment, simply because its hearing is used for so many purposes. By contrast, in the study of hearing in bats, we know precisely what the animals are listening to: they are listening to themselves. The auditory system of bats is largely dedicated to analysing the echoes of their own cries as part of the sonar system by which they find their way around (see Chapter 6). Knowing this central fact, the physiological properties of nerve cells in the auditory system are readily correlated with their behavioural function in the intact animal.

Whether or not the system under study is a dedicated one, it obviously makes the neuroethologist's task easier if the absolute number of nerve cells involved is small. Unfortunately, most of the higher vertebrates have very large numbers of nerve cells in even the smallest subsection of their central nervous systems. By and large, therefore, neuroethologists have looked to the lower vertebrates and the invertebrates for suitable study material. Among the invertebrates, the arthropods show behaviour that is complex enough to be interesting yet they also show a remarkable economy in the number of nerve cells involved. Whereas mammals may employ hundreds of nerve cells to excite a single muscle, for example, arthropods usually make do with no more than half a dozen (see Chapter 7). It does not necessarily follow that, because the number of nerve cells involved is smaller, the neural principles of operation will be simpler. But working with a smaller number of nerve cells does increase the chances of discovering the principles in the first place.

1.3 Neural implications of ethological results

The behaviour of an animal is to a large extent the product of activity in its nervous system. The patterns of behaviour that are recognised in ethological studies must therefore reflect the underlying organisation of the nervous system. In the case of the elementary components of behaviour studied by the early ethologists, this correspondence may be fairly close. Consequently, a careful study of behaviour patterns at the level of the intact organism will often produce results that provide valuable clues about the underlying neural organisation.

Consider the classic case of the egg-retrieval behaviour found in many ground-nesting birds, which was first studied in the greylag goose (genus *Anser*) by Lorenz and Tinbergen in the 1930s. A nesting goose employs a stereotyped sequence of movements to retrieve an egg that has become displaced from the nest. The bird leans out of the nest, places its beak beyond the egg, and then draws the beak back towards its chest so that the egg is rolled back into the nest. Superimposed on this movement towards the chest are little side-to-side movements of the beak, which serve to keep the egg in place. This sequence of movements is used by all members of the species for egg retrieval; none uses an alternative method. Indeed, a very similar pattern of movement is found in other birds, such as the herring gull (*Larus*), on which many tests have been carried out (Fig. 1.1). Stereotyped movements of this kind were originally called **fixed action patterns**; nowadays, more general terms like **motor pattern** are used instead by most ethologists.

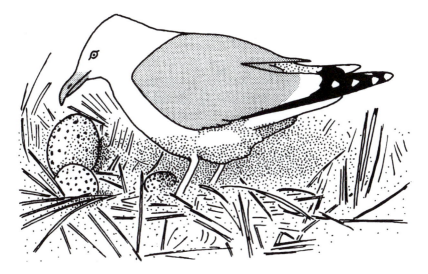

Figure 1.1 Egg retrieval in the herring gull (*Larus*): an incubating gull will retrieve an egg that has become displaced from the nest, using a stereotyped pattern of movement. Here, the retrieval response is being used to test what the gull perceives to be an egg. Two different models, both of which differ considerably from the real egg in the nest, are placed on the rim of the nest to compare their effectiveness in eliciting the retrieval response. (Redrawn after Baerends & Drent, 1982.)

It was noticed that many such motor patterns seem to occur in response to specific stimulus situations in the natural environment. During the 1930s, ethologists developed the technique of using models, in which one feature at a time could easily be varied, to find out what features of a situation are important in triggering an animal's response. Lorenz and Tinbergen found that the greylag geese would retrieve wooden models painted to resemble natural eggs. The goose would still retrieve the models when they were made the wrong shape, such as cubes or cylinders, or when they were made the right shape but the wrong size, including models that were much larger than a normal egg. It was evident from these results, and many others, that only certain features of the natural stimulus are needed to produce a response. These essential features were called **sign stimuli** or, where they were found in the context of social behaviour, **social releasers**.

Ethologists rightly sought to account for the fact that animals often respond to only a small selection of the available stimuli by postulating neural mechanisms in the responding animal. Response selectivity might

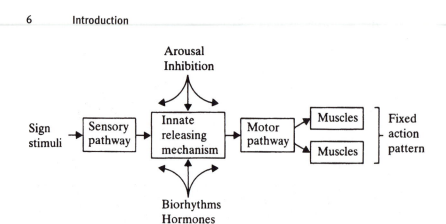

Figure 1.2 A flow diagram showing early ethological concepts of the mechanisms involved in a simple behaviour pattern such as egg retrieval. (Redrawn after Shepherd, 1983.)

be due partly to the capacities of the sense organs, but it was already known that an animal may respond to a specific sensory cue in one behavioural context and not in another. Hence, the occurrence of sign stimuli must also be due to stimulus selection by more centrally located mechanisms processing the information received from the sense organs. The term **releasing mechanism** was coined for this central processing and, because it was assumed to develop independently of experience with the sign stimuli, the adjective innate was attached to it, giving innate releasing mechanism (IRM). The adjective innate is not much used by modern ethologists, but the term releasing mechanism continues to call attention to an important phenomenon of behaviour.

The way in which the various components might interact to produce a behaviour pattern is illustrated in Fig. 1.2, which represents the results of the early ethological period. In egg retrieval, the visual stimuli from around the nest are passed from the sense organs along a neural pathway to the central nervous system, where the releasing mechanism responds to the sign stimuli that indicate 'egg'. This central mechanism then releases or triggers activity in the motor regions of the nervous system that generate the fixed action pattern for retrieval. This sequence is not invariable in its operation but is enhanced or prevented by other factors. Thus, the releasing mechanism is inhibited in the short term (arrows from above in Fig. 1.2) when the bird is away from the nest foraging or escaping from a predator, and in the long term (arrows from below) retrieval cannot be elicited outside the breeding season, which is controlled by reproductive hormones.

Further insight into this phenomenon has been made possible by the detailed studies of egg retrieval in the herring gull carried out by Baerends and his colleagues (Baerends & Drent, 1982; Baerends, 1985), who placed two egg models side by side on the rim of the nest and then watched from a hide to see which of the models the gull retrieved first. Thousands of these tests were made, carefully varying only one feature at a time, in order to determine what the gulls' preferences were. It was found that the gulls preferred larger eggs to smaller ones, green eggs to any other colour, speckled eggs to uniformly coloured ones, strongly contrasting speckles to weakly contrasting ones, and natural egg shapes to abnormal ones. This last preference was not nearly as strong as might have been expected, and a cylindrical model was almost as effective as an egg-shaped model of the same size and colour.

These results show that the gulls do, indeed, respond selectively to a limited number of stimuli, which match a real gull's egg only in a rough way. It is not even necessary for all the stimuli to be present for a response to occur. The stimuli that are present add together independently to determine the overall effectiveness of an egg model in producing a response. For instance, a smaller green egg will be as effective as a larger brown egg; if speckling is then added to the green egg, it will become more effective than the larger brown egg. One consequence of this property is that models can be made more effective than the real object they represent. A gull will retrieve a model 50 per cent larger than normal, green and with black speckling in preference to one of its own eggs; such a model is what ethologists call a supernormal stimulus.

The experiments with models show that this releasing mechanism involves perception of a number of simple visual cues, which add together quantitatively to determine the degree of 'egginess' as far as the gull is concerned. Clearly, these properties reflect the way in which visual perception occurs in the gull's nervous system, and the flow diagram shown in Fig. 1.3 tries to incorporate this. The response to a limited number of simple cues may well reflect the occurrence in the early stages of the visual system of units that respond selectively to visual cues such as colour, contrast, edges and shapes (represented as selectors, S_1 to S_9, in Fig. 1.3). The way in which the separate cues add together suggests the presence of a more central unit

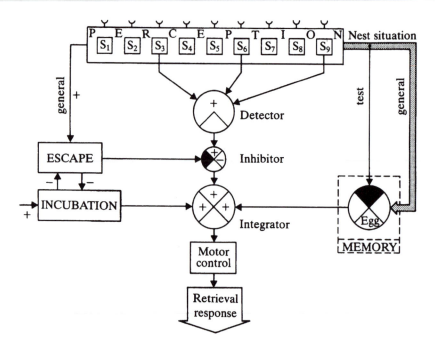

Figure 1.3 Releasing mechanism for egg retrieval in the herring gull: a flow diagram based on experiments with egg models. The boxes represent major systems or operations and the circles indicate sites where summation of inputs occurs. Visual perception (top) is represented as a series of selectors (S_1 to S_9) that respond to particular features of the stimulus. Some of these feed on to a specific detector for egg recognition, which in turn feeds on to the motor control for egg retrieval. This response is maintained during the period of incubation but may be overridden by other factors such as the need to escape (left) or the bird's memory based on experience with real eggs (right). (Redrawn after Baerends, 1985.)

that combines information from a specific set of selectors so as to act as a detector for specific objects in the environment, in this case an 'egg detector'. Units that correspond closely with this description are found widely in the visual systems of both vertebrates and invertebrates, as shown in the following example of prey detection in toads (see also Chapter 5). It is easy to see how these units could be excited more strongly by a supernormal combination of stimuli than by the natural combination.

1.4 Sign stimuli in amphibians

The way in which frogs and toads recognise their prey provides another example of a releasing mechanism. In this case, the ethological results are even more compelling because they have been combined with a neurophysiological study of the same system. This combined approach clearly shows how the selective properties of nerve cells (neurons) are involved in releasing particular patterns of behaviour (Ewert, 1985, 1987).

In the visual world of a frog or toad, just a few, simple criteria serve to categorise moving objects as prey, enemy or lover. Once the visual system has placed a given object in one of these categories, the animal reacts accordingly. These reactions can be used to analyse the criteria involved in prey recognition because the animals are readily deceived by small cardboard models moving in front of them. A special study of prey detection has been made in the common European toad (genus *Bufo*), using such models to analyse the behavioural responses of the intact animal and the responses of specific classes of neuron in the visual system. The natural prey of *Bufo* consists of small animals such as beetles, earthworms and millipedes. If one of these animals appears in its peripheral visual field, the toad responds by turning its head and/or body so as to bring the animal into the frontal visual field. The toad then walks towards the prey in order to capture it.

The sign stimuli, by which the prey is recognised, can be analysed quantitatively in the laboratory. A hungry toad is confined in a glass vessel, from which it can see a cardboard model circling around (Fig. 1.4 *a*). If the toad interprets the model as a prey animal, it tries to bring it into the frontal visual field, and in doing so turns around jerkily after the moving model. The number of orientating turns per minute elicited by a given model, compared to the number elicited by others, can therefore be taken as a measure of the resemblance between that model and prey, from the toad's point of view.

In this experimental situation, the toad is not much impressed by a small 2.5×2.5 mm model, which elicits only a few orientating movements. However, the stepwise elongation of this shape in the horizontal dimension (Fig. 1.4*b*, shape x) greatly increases its releasing value. That is to say, elongation of the model in the direction of movement increases its resemblance to prey, up to a certain limit, and this long, small stripe has been called the

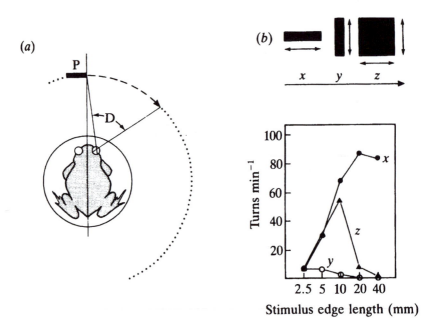

Figure 1.4 Analysis of prey recognition in the toad (*Bufo*). (*a*) The experimental set-up, with the toad confined in a glass vessel and a prey model (P) circling around it. The toad turns to follow the model when it has moved through a sufficient angle, the effective displacement (D). (*b*) The response of the toad to moving models of three shapes (*x*, *y*, *z*) as these are enlarged in one dimension (shapes *x*, *y*) or two dimensions (shape *z*). The toad's response is measured by the number of times it turns to follow the model in 1 min. (Redrawn after Ewert, 1980, 1983.)

worm configuration. If the small, square shape is elongated in the vertical dimension (Fig. 1.4*b*, shape *y*), its releasing value decreases to zero. In fact, the toad often interprets it as a threat and freezes in a defensive posture. This shape has been called the antiworm configuration. If both dimensions of the model are lengthened equally, so that the toad is presented with squares of increasing size (Fig. 1.4*b*, shape *z*), the prey-catching activity initially increases but then declines rapidly to zero. This is probably the result of non-linear summation of the horizontal (worm) and vertical (antiworm) edges.

The toad's ability to distinguish between worm and antiworm does not vary with other stimulus parameters, such as the colour of the model or its velocity of movement. It is also independent of the direction in which the stimulus traverses the toad's visual field. If the models are moved past the toad in a vertical direction, then the vertical stripe elicits prey catching and the horizontal stripe elicits no response or a defensive posture. Thus, the worm/antiworm distinction is based on the combination of just two stimulus parameters: the elongation of the object in relation to its direction of movement. These parameters, then, are the sign stimuli that release prey-catching behaviour in a hungry toad, and it is obvious that they correspond only very approximately to a real worm. Nevertheless, they will normally enable a toad to distinguish correctly between potential prey and inedible objects in its natural environment.

1.5 Neuroethology of a releasing mechanism

As with other vertebrates, early visual processing in amphibians takes place in the neuronal circuits of the retina. The neurons of the vertebrate retina (see Fig. 2.4, p. 25, and Box 5.1, p. 107) are arranged in a way that provides for both lateral interaction and through transmission. The through route of the visual pathway is made up of receptors, bipolar cells and ganglion cells.

Recording with microelectrodes shows that the receptors respond in a simple way to changes in the intensity of light that falls on them as part of the image formed by the eye. These responses are mirrored by the bipolar cells, each of which receives input from several receptors. In turn, the ganglion cells each pool input received from a large number of bipolar cells. The ganglion cells are able to respond to more complex features of the image because of the way information from the receptors is processed and combined on its way to the ganglion cells. Ganglion cells can be divided into several different classes on the basis of the type of feature to which each is most sensitive. Some are most sensitive to objects of a given angular size, others to objects moving in a particular direction or to the difference in brightness between adjacent areas of the image. This information is passed along the optic nerve to the brain (Fig. 1.5), where these basic parameters are used to distinguish between mate, prey and predator.

The majority of retinal ganglion cells are connected to the optic tectum, a specialised region of the midbrain visible as a large bulge on either side (Fig. 1.5). A smaller number of ganglion cells are connected to the thalamus, which is the most prominent part of the posterior forebrain, and to the pretectal areas of the midbrain. These connections are spread out in an

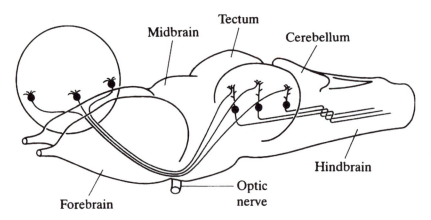

Figure 1.5 The layout of the main visual pathways concerned in prey detection in the brain of an anuran amphibian. The axons of most ganglion cells travel from the eye to the optic tectum on the opposite (contralateral) side of the brain, via the optic nerve (other cranial nerves are not shown). Feature-detecting neurons of the optic tectum send their axons to the motor regions of the contralateral hindbrain.

orderly manner in the superficial layer of the optic tectum, with each ganglion cell keeping the same relative position with respect to its neighbours that it has in the retina.

The responses of the tectal neurons can be recorded by probing the deeper layers of the optic tectum with a microelectrode (Fig. 1.6). Like the ganglion cells, the neurons of the thalamus and tectum can be divided into different classes according to their patterns of response. Of the thalamic and tectal neurons that have been investigated, at least three classes show differing responses to moving stimuli of worm and antiworm configurations. The thalamic Class TH3 neurons respond best to squares; stripes with the antiworm configuration elicit a lesser response, and the worm configuration elicits the least response of all (Fig. 1.6b). In the optic tectum, the Class T5(1) neurons also respond best to squares, but when tested with stripes, they prefer the worm to the antiworm configuration (Fig. 1.6c). Another class of tested cells, the Class T5(2) neurons, distinguish much more clearly between the worm and antiworm configurations, with the worm configuration eliciting the greatest response, the squares a lesser response, and the antiworm by far the least response (Fig. 1.6d). Among all the neurons tested so far, the response pattern of the T5(2) neurons shows

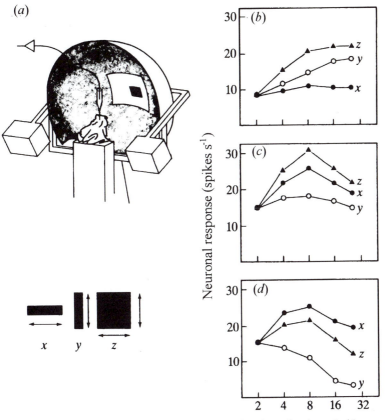

Figure 1.6 (*a*) Set-up for recording the responses of neurons in the brain to moving visual stimuli. The toad is held in a fixed position and its brain is probed with a microelectrode for recording the spikes in single neurons. Each stimulus is moved in front of the toad by means of the perimeter device. (*b*) The response of thalamic Class TH3 neurons to increasing angular size of the same three shapes (*x*, *y*, *z*) used in the behavioural tests (Fig. 1.4). (*c*) The response of tectal Class T5(1) neurons to the same three shapes. (*d*) The response of tectal Class T5(2) neurons to the same shapes. (*a* redrawn after Ewert, 1985; *b–d* redrawn after Ewert, 1980.)

the best correlation with the sign stimuli for prey-catching behaviour (cf. Fig 1.6*d* and Fig. 1.4*b*).

It is evident from Fig. 1.6 that the responses of the Class T5(2) neurons could be accounted for if they receive excitatory input from Class T5(1) neurons and inhibitory input from Class TH3 neurons. The fairly strong response to the worm configuration in Class T5(1) neurons would be minimally inhibited by the poor response to it in the Class TH3 neurons, resulting in a strong response in the Class T5(2) neurons. Similarly, the poor response to the antiworm configuration in Class T5(1) would interact with the moderate response in Class TH3 to give a very poor response in Class T5(2).

This possibility has been tested by removing the input from the Class TH3 neurons, which can be accomplished by severing the pathway that is known to run from the thalamus to the optic tectum. Whether this lesion is done permanently by microsurgery or temporarily by local application of a neurotoxin, the effect on Class T5(2) neurons is dramatic. The responsiveness of these neurons to all visual stimuli is increased and selectivity is lost, with the neurons responding best to squares and failing to distinguish clearly between stripes in worm and antiworm configurations. This shows that the normal selective response of the Class T5(2) neurons is dependent on inhibition from thalamic neurons, including the Class TH3 neurons.

When a toad with a pretectal lesion is allowed to recover from surgery and is tested behaviourally, its responses closely parallel those of the T5(2) neurons: the operated animal responds vigorously to all shapes, preferring squares and failing to distinguish clearly between worm and antiworm configurations of stripes. Such a close correspondence between the responses of the Class T5(2) neurons and of the whole animal suggests that these neurons are directly involved in prey detection and hence in releasing prey-catching activity. This is confirmed by means of a small telemetry system that enables the experimenter to record from and stimulate single neurons in the optic tectum of a freely moving toad. Recordings made with this system show that activity of Class T5(2) neurons precedes and continues during the orientation of the toad towards the prey. Having recorded in detail from a particular T5(2) neuron, it can then be stimulated by passing a tiny current through the microelectrode, and this consistently elicits orientating movements that are directed to the appropriate part of the visual field.

If they are involved in prey detection in this way, one would expect the Class T5(2) neurons to be connected, directly or indirectly, with the motor circuits in the hindbrain of the toad. Various histological methods demonstrate that a number of the connections arriving in the motor regions on one side of the hindbrain do come from the contralateral optic tectum (see

Fig. 1.5). That these include the Class T5(2) neurons is confirmed by physiological methods. Localised stimulation of the appropriate neural tract in the hindbrain sends signals travelling back up to the optic tectum, where they can be recorded in individual T5(2) neurons with a microelectrode.

Thus, the Class T5(2) neurons provide an excellent example of specific brain cells that are involved in releasing a simple, important behaviour pattern. On the basis of the stimulus parameters selected by the retinal ganglion cells, the neurons of the optic tectum are able to respond to specific parameter combinations that carry information relevant to the toad's way of life. The specific combination of visual parameters to which the T5(2) neurons respond carries information enabling the toad to distinguish between its natural prey and inedible objects. These response properties of the T5(2) neurons do not identify worms or beetles, but rather they underlie the sign stimuli that elicit prey-catching behaviour in a hungry toad.

1.6 Control theory and nervous systems

In the attempt to pursue analysis of the mechanisms of behaviour to the neural level, most of the concepts used are drawn either from general neurobiology or from ethology itself. An additional strand of thought that has made a useful contribution is cybernetics or control systems theory, developed to provide a formal analysis of human control systems. It is often useful to treat the nervous system of a behaving animal, or some part of its nervous system, as a control system through which there is an orderly flow of information, with definite input and output elements.

This approach has provided a helpful terminology and way of thinking about the mechanisms of behaviour. With the advent of powerful computers, it has also become possible to use this approach for constructing models that mimic the interactions among groups of neurons. When used carefully, this provides a means of testing whether circuits work in the way that is predicted from physiological recordings. This is particularly useful when large numbers of neurons are involved because it is rarely possible to monitor activity in many neurons simultaneously.

An example of a useful concept from control theory is that of **feedback**; this takes place when an event or process has consequences which affect the occurrence of that event or process. In control systems, feedback is usually negative, which means that action is taken in response to a

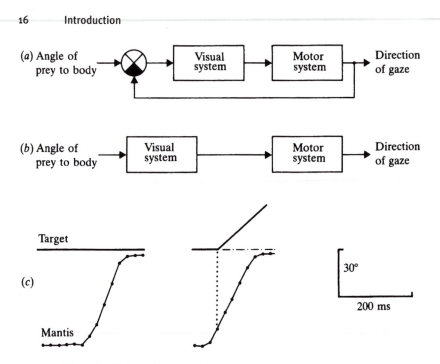

Figure 1.7 Orientation to prey in a mantis (*Tenodera*), illustrating movements with and without feedback. (*a*) A flow diagram of visual tracking of prey with feedback. (*b*) A flow diagram of rapid visual location of prey without feedback. In (*a*) and (*b*) the boxes represent major systems or operations and the circles indicate sites where summation of inputs occurs. (*c*) Rapid location of prey, as in (*b*), analysed from videotape, showing orientation towards a stationary target (left) and towards a target that moves off at a constant angular velocity after the mantis has started to turn (right). (*c* redrawn after Rossel, 1980.)

disturbance so as to correct that disturbance. This usually involves a special mechanism like a thermostat, by means of which the output of a system is fed back to regulate the input. Many behaviour patterns also have this self-regulatory character. In egg retrieval, the movement of the beak towards the chest seems not to involve feedback because it is so stereotyped and because it continues to completion even if the egg is removed during the movement. However, the side-to-side movements that keep the egg centred on the beak certainly appear to involve feedback of some kind; they tend to disappear if the bird is retrieving a cylindrical model, which rolls smoothly and so does not need centring.

The distinction between movements that do and do not involve feedback is made clear by the visual orientation to prey in a praying mantis. The insect follows potential prey with movements of its head or body so as to keep the prey in the centre of its line of vision. In this behaviour pattern, visual information triggers a movement of the praying mantis, and this results in an altered visual input, which in turn influences the subsequent movement. Hence, the flow of information forms a **closed loop**, with output feeding back to the input (Fig. 1.7*a*). However, a different situation obtains when the mantis first locates the prey. As soon as a suitable object appears in the visual field, the mantis turns towards it with a rapid movement that is not influenced by feedback from the visual system. Even if the object is experimentally removed during the turn, the mantis still continues turning until it faces the place where the object originally was. Hence, in this case, the information flow forms an **open loop**, without feedback (Fig. 1.7*b, c*).

In common with ethological concepts considered above, such concepts from control theory do not in themselves provide an explanation in terms of underlying mechanisms. Rather, they are, in computer terminology, 'software' explanations that specify the job done and the relations between the different components of a behaviour pattern. For a full understanding of the mechanisms of behaviour, these concepts obviously need to be coupled with a detailed analysis of the underlying neural 'hardware'. Once a neural analysis is accomplished, or at least is underway, then the 'software' concepts come into their own as a vehicle for showing how the neural 'hardware' is organised so as to generate a given behaviour pattern. To take a simple example, flow diagrams such as that in Fig. 1.3 have been borrowed from control theory as a vehicle for summarising behavioural mechanisms. If such a diagram is based on known neural components, whose physiological properties have actually been studied, rather than on hypothetical 'black boxes', then the diagram becomes a truly effective way of summarising the link between the nervous system and behaviour (see, for example, Fig. 3.7, p. 60).

1.7 Conclusions

The biological study of animal behaviour was put on a sound footing by the early ethologists in the middle of the twentieth century. These ethologists pioneered techniques for investigating the natural behaviour of animals, including the use of simple experiments in conjunction with field observations. In the course of this work, they developed a number of key concepts that have helped to guide efforts to understand the mechanisms of behaviour.

One such concept is that of the motor pattern, which is a relatively stereotyped sequence of movements and is easily recognised as part of an animal's ongoing behaviour. The recognition of these motor patterns in the animal's natural behaviour clearly implies that there is a corresponding pattern in the underlying organisation of the animal's nervous system, which generates these movements. Another concept is that of the releasing mechanism, which may be envisaged as a kind of neural filter tuned to recognise specific sign stimuli in the environment. The nervous system must be organised so as to sort out different stimuli and to make decisions about which motor pattern to put into action at any one time. The egg retrieval response in nesting birds provides a good example of the development of these two concepts.

The way that particular nerve cells can be identified as playing specific roles in the control of behaviour is well illustrated by the T5(2) neurons in the brains of toads. The close correspondence between the responses of these neurons and the sign stimuli for prey catching in the toad (*Bufo*) suggests that these cells are part of the natural releasing mechanism for prey-catching behaviour. This conclusion is backed up by lesion experiments, which alter the responses of these nerve cells and the behaviour of the intact animal in similar ways.

Much recent research in neuroethology is aimed at understanding how nerve cells are organised into circuits that perform specific functions in behaviour, such as filtering out sign stimuli or generating a particular motor pattern. This work involves tracing the flow of signals from one nerve cell to the next, and is most easily done where particular nerve cells can be uniquely identified. An essential first step along this path to understanding how nervous systems underlie behaviour is to examine the relevant properties of the fundamental units of the nervous system, the nerve cells.

Further reading

Alcock, J. (1998). *Animal Behavior, an Evolutionary Approach*, 6th edn. Sunderland, MA: Sinauer Associates. A useful textbook, which includes some neuroethology, setting it nicely in the context of the varied questions that can be asked about the mechanisms and evolution of behaviour.

Manning, A. & Dawkins, M.S. (1998). *An Introduction to Animal Behaviour*, 5th edn. Cambridge: Cambridge University Press.
Another good textbook, which gives neuroethology its place among the diversity of subject matter and levels of analysis included within the study of behaviour.

Ewert, J.-P. (1980). *Neuroethology*. Berlin: Springer-Verlag. One of the first textbooks treating neuroethology as a distinct area of study, it remains useful for its expert treatment of prey recognition in toads.

Ewert, J.-P. (1985). Concepts in vertebrate neuroethology. *Anim Behav* **33**, 1–29.
This brief essay clearly explains the ideas about neural filtering that have emerged from research on prey recognition in toads, using both ethological and neurophysiological methods.

The International Society for Neuroethology provides a Web site with many links to interesting aspects of this branch of biology: http://www.neuro-bio.arizona.edu/isn/

2 Nerve cells

2.1 Basic organisation of nerve cells

The very title of this chapter would have been contentious in the nineteenth century, when detailed scientific study of the nervous system got underway. For it was then not generally agreed that the nervous system is composed of many individual nerve cells. This was mainly due to the fact that nerve cells are difficult to visualise with routine histological methods. Many cells are packed tightly together in nervous tissue (there are 100 000 nerve cells in 1 mm³ of human brain) and they give off fine, branched processes, so that it is almost impossible to determine the limits of a single cell. Many scientists therefore believed that nerve cells were fused together in a continuous network of branched processes, rather like the capillary beds that link small arteries and veins.

The technique that was most powerful in challenging this view was silver staining, first discovered by Camillo Golgi in 1873 and developed by others, particularly Santiago Ramon y Cajal from 1888 onwards. Ramon y Cajal examined many parts of the nervous system in a wide range of animal species. He realised that the special feature of silver staining is that it only stains a small percentage of cells in a piece of tissue but it stains them in their entirety, so that the structure of an individual nerve cell can be described (Fig. 2.1). Nowadays, single neurons are often stained by the intracellular injection of dye through a microelectrode, or by the use of methods that recognise chemicals characteristically found in particular neurons. In 1891, an anatomist named Wilhelm Waldeyer reviewed all the evidence from microscopical and developmental studies and concluded that the nervous system, like other organ systems, is indeed composed of discrete cells. He suggested the term **neuron** for the nerve cell.

Final confirmation that the nervous system is composed of discrete cells

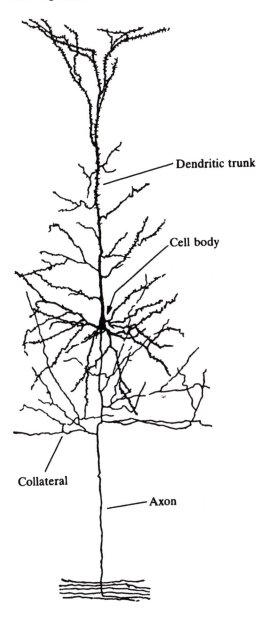

Figure 2.1 Drawing of a pyramidal cell from a mammalian brain, an example of a complex nerve cell. This is a single cell, stained by the Golgi silver method. At the bottom of the figure, the axon is shown joining a bundle of axons from similar cells. The cell body is about 20 μm (0.02 mm) across. (From Ramon y Cajal, 1911.)

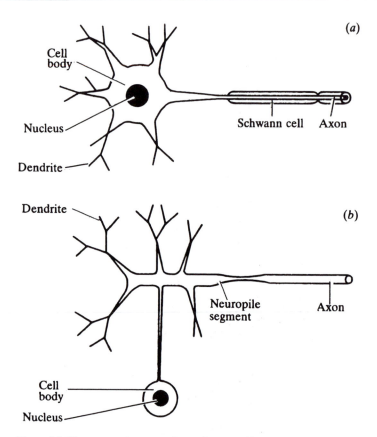

Figure 2.2 Diagrammatic comparison of nerve cell structure in (*a*) vertebrates and (*b*) arthropods, illustrated by motor neurons.

was provided in the 1950s by studies with the electron microscope, which revealed that a gap separates the plasma membrane of adjacent nerve cells. These studies also showed that a nerve cell contains an assemblage of intracellular organelles, such as endoplasmic reticulum and mitochondria, just like any other animal cell. These organelles are required to carry out the usual metabolic functions of an animal cell. At the same time, nerve cells are specialised in their anatomy and physiology to carry out the particular function of processing information.

The specialised anatomy of nerve cells can be seen in Fig. 2.2, which diagrammatically compares a vertebrate neuron with that of an arthropod. As in other cells, the main intracellular organelles are gathered around the nucleus, and the region of the neuron that includes the nucleus is called the

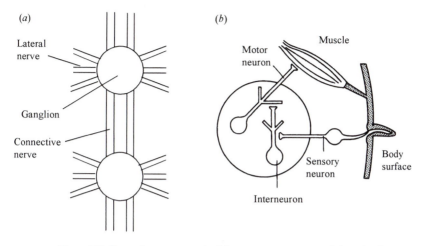

(a)

Lateral nerve

Ganglion

Connective nerve

(b)

Muscle

Motor neuron

Sensory neuron

Body surface

Interneuron

Figure 2.3 General arrangement of the nervous system and three basic categories of neuron, illustrated by schematic diagrams based on an arthropod. (*a*) Two segmental ganglia, showing lateral and connective nerves. (*b*) One ganglion with one sensory neuron, one motor neuron and an interneuron.

cell body. Neurons involved in long-distance signalling have a long, cable-like process, the **axon**, which is devoid of most intracellular organelles. Other processes are shorter and often branch repeatedly; these are usually called **dendrites**. In most vertebrate neurons, the dendrites and the axon emerge from the cell body, but in many invertebrate neurons the cell body is separated from the branching region and lies at the end of a thin process, or neurite. In many neurons concerned with local signalling the axon is short, and may even be absent altogether. Neurons with no axon are called amacrine cells and all their processes are dendrites.

The precise arrangement of axon and dendrites varies greatly among neurons, and different types of neurons can be identified from their particular branching pattern. Sometimes the pattern can be very complex, as in the case of the pyramidal cell in Fig. 2.1. Here, the many dendrites arise not only from the cell body but also from a stout dendritic trunk. The axon is relatively thin, and it gives off axonal branches called collaterals fairly close to the cell body.

In terms of function, neurons can be grouped into three basic categories (Fig. 2.3*b*). Neurons with specialised endings that respond to energy from the environment are called **sensory neurons**. Neurons that have axons terminating on muscle fibres are called **motor neurons**. All other neurons are

interneurons. Some interneurons connect relatively distant parts of the nervous system and have long axons, but many are amacrine cells or have short axons, and these are often referred to as local interneurons. As well as neurons, the nervous system contains cells that surround and support the neurons, and these are called **glial cells** (glia = 'glue').

Neurons tend to be clustered together in structures called **ganglia** (Fig. 2.3). In clearly segmented animals, such as arthropods or annelid worms, there is one ganglion for each body segment, although in many species the ganglia of adjacent segments are fused together. A segmental ganglion contains the motor neurons that control the muscles of its segment (Fig. 2.3a), and axons from the sense organs of the segment run into it along lateral nerves. Ganglia of adjacent segments are linked by relatively stout nerves called **connectives**, which consist of bundles of axons that almost all belong to interneurons. Ganglia occur in many non-segmented animals, such as molluscs, and some aggregations of neurons in the vertebrate central nervous system are also called ganglia. In most invertebrate ganglia, the cell bodies of motor neurons and interneurons occur in a thin rind on the periphery of the ganglion, and the core that contains axons and dendrites is called the **neuropile**.

Although neurons are separate entities, they are connected together in specific ways to form circuits, in which signals travel from cell to cell along particular pathways. In the simplest pathways, signals pass from sensory neurons straight to motor neurons, but most pathways involve several interneurons. During the operation of a nervous system, signals pass sequentially from place to place in individual neurons, and from neuron to neuron in a circuit. As a general rule, the dendrites of a neuron are where it receives signals, and the axon terminals are where the neuron transmits signals onwards to other neurons or to muscle cells. A neuron adds together signals that it receives from other neurons and usually converts them into a new signal that it may pass on. The process of combining inputs together to produce a new output signal is called **integration**.

Ramon y Cajal deduced from his anatomical studies that neurons are connected into functional circuits and clearly understood that signals flow from cell to cell. One part of the nervous system he studied intensively was the vertebrate retina (Fig. 2.4). He showed that the very short axons of sensory neurons (rods and cones) make contact with the dendrites of bipolar cells, which in turn have axons that contact the dendrites of retinal

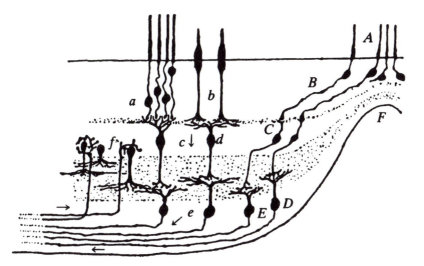

Figure 2.4 The pattern of connections in the primate retina, as revealed by
Golgi silver staining. The diagram includes sensory neurons (rods and
cones, a, b, B, A), bipolar cells (c, d, C) and the retinal ganglion cells (e, E,
D). F indicates the fovea, where bipolar and retinal ganglion cells are displaced from their receptors but the normal pattern of connections is maintained. Some amacrine cells (f) are included on the left, but horizontal cells,
at the sensory neuron–bipolar cell junctions, are omitted. (From Ramon y
Cajal, 1911.)

ganglion cells. The retinal ganglion cells have long axons with fine,
branched terminals in the optic region of the brain. Thus, information
about vision must flow through this sequence of connections from the
receptors through bipolar and retinal ganglion cells to visual centres in
the brain. In such a sequence, neurons are often referred to by their order
in the chain, so sensory neurons are first order, bipolar cells second order,
retinal ganglion cells are third order and cells in the brain are higher order.

Signals do not just flow sequentially from cells of one order to the next,
however, but also flow laterally. Ramon y Cajal indicated this on the left of
his diagram (Fig. 2.4) by showing retinal amacrine cells that contact the
dendrites of retinal ganglion cells and axon terminals of bipolar cells. The
amacrine cells serve to modify signals as they pass from bipolar cells to
retinal ganglion cells, and may allow activity in one bipolar cell to modify
the information that its neighbouring bipolar cells transmit to retinal ganglion cells. In addition, there is another layer of horizontally oriented cells

in the retina, the horizontal cells, that lies at the level of the junctions between sensory neurons and bipolar cells.

2.2 Neuron physiology and action potentials

Neurons are specialised in their physiology to receive, sort out and pass on information. The signals that neurons deal with involve small changes in the electrical voltage between the inside and outside of the cell, and integration is the process by which these voltage changes are combined together to determine the neuron's output signal. This is essentially how neurons make decisions.

The most conspicuous of these voltage changes are **action potentials**, which are the signals that neurons use to transmit information along long axons. Each action potential lasts slightly less than a millisecond (1 ms) at a particular location along the axon and travels along the axon at a speed that varies from less than 1 m/s to nearly 100 m/s depending on the girth of the axon. Action potentials may be recorded from outside an axon by using fine wires as electrodes, as shown in Fig. 2.5a. Although the voltage change between the inside and outside of the neuron is about a tenth of a volt, the signal that the electrodes outside the cell pick up is much smaller, so considerable amplification is needed to display and measure the action potential. Each action potential is conducted rapidly along the axon and passes the two electrodes in succession. As it passes the first electrode, the latter will become positive with respect to the second electrode, and as the action potential passes the second electrode, the situation is rapidly reversed. Consequently, the output signal that the amplifier delivers has an S-shaped waveform when displayed on the oscilloscope (Fig. 2.5b). On a slower time scale, action potentials appear as stick-like departures from the baseline, and because of this appearance they are commonly called **spikes**.

A more precise method of recording a spike at one location along an axon is to use an intracellular electrode. This consists of a glass capillary tube that is drawn out to a very fine point and filled with a conducting salt solution such as potassium acetate. The electrode is connected to an amplifier by way of a silver wire placed into the electrode (Fig. 2.6a). When the tip of the electrode is inserted through the membrane of the cell, the salt solution is in electrical contact with the inside of the cell, and the signal recorded measures the difference in electrical potential, or voltage, between the

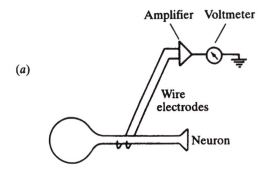

(a)

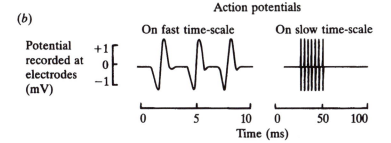

(b)

Figure 2.5 Activity in the axon of a neuron, recorded with extracellular electrodes. (a) The recording method shown schematically. (b) The typical appearance of action potentials when recorded by this method and displayed on an oscilloscope. The oscilloscope is a sensitive voltmeter, and displays changes in voltage recorded with time.

inside and the outside of the axon. The voltage recorded by an intracellular electrode is called the **membrane potential**. When the tip of the electrode first enters the cytoplasm, it usually records a voltage of 60–80 mV across the membrane, with the inside of the cell negative with respect to the outside. This is called the **resting potential**.

The cell membrane of a resting cell is thus polarised: there is a standing electrical voltage across the cell membrane. A reduction in electrical potential across the membrane, bringing the intracellular voltage closer to the extracellular, is called a **depolarisation**. Most depolarising events are **excitatory** because they increase the likelihood that the neuron will generate an action potential. An increase in the membrane potential from the resting value is called a **hyperpolarisation**, and the effect of this is **inhibitory** as it counteracts any depolarisation. Much of the process of integration involves

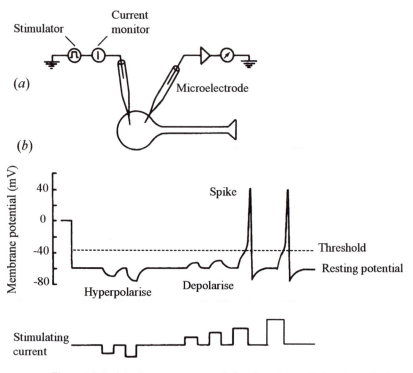

Figure 2.6 Activity in a neuron recorded with an intracellular electrode. In this method (*a*), one electrode is used to inject current pulses of different magnitudes into the neuron and a second is used to record the voltage, or potential, across the membrane. In (*b*), the typical appearance on an oscilloscope of responses by a neuron to pulses of current is shown.

an interplay between depolarising and hyperpolarising membrane potentials.

A second intracellular electrode may be used to stimulate the neuron with pulses of electrical current, as shown in Fig. 2.6. Negative pulses of current hyperpolarise the neuron, producing downward deflections in the voltage recording trace. Positive pulses of current depolarise the neuron, causing upward deflections in the voltage trace. If the depolarising current pulses are small, the neuron's response is passive, and the size of the voltage change is proportional to the size of the current stimulus. However, when the depolarising potential reaches a critical **threshold** value, the neuron responds actively by producing a spike. Membrane potential changes rapidly, reaching a peak when the inside of the neuron is about 40 mV positive in relation to the outside, and then returns to its resting level, often after a transient hyperpolarisation. A second spike cannot be initiated until a short time after the first one is complete, this time being called the **refractory period.**

When a spike occurs at one location, it depolarises the membrane for some distance away. This depolarisation acts as the stimulus for new spikes, but, because of the refractory period, only in the length of axon that has not recently produced a spike. A spike is therefore a stereotyped event, in which the membrane potential swings rapidly between resting potential to about 40 mV positive and back. The amplitudes of spikes in extracellular recordings appear to vary from axon to axon, but this is because the size of the extracellular signals picked up by the electrodes depends on the diameters of the axons and how far away the axons are from the electrodes.

A difference in the voltage inside and outside a cell is the consequence of two features of cell physiology. The first is that charged ions are distributed unevenly between the cytoplasm and extracellular fluid, and the second is that the membrane contains pores that, when open, allow particular ions to pass through. The resting potential arises because potassium ions are more concentrated in the cell's cytoplasm than in the extracellular fluid, and pores that are open in the resting cell membrane allow these ions (but not others such as sodium or chloride) to pass through readily. Driven by the concentration gradient, potassium tends to flow out of the cell. As each potassium ion flows out of the cell, however, it carries with it a positive charge, making it successively more difficult for subsequent potassium ions to leave. A balance is established between the concentration gradient that tends to push potassium outwards and the electrical gradient that tends to retain it inside the cell. The electrical gradient at the balance point is the resting potential.

The pores that allow ions to pass through the membrane are usually formed by specialised protein molecules that aggregate in particular formations. These proteins and their associated pore are called a **channel**. The resting membrane potential is therefore determined by the action of one kind of potassium channel. During a spike, two other types of channel, a sodium channel and another type of potassium channel, come into play. These two channel types have pores that are closed in the resting cell, but tend to open when the neuron is depolarised to the threshold voltage and beyond. Electrical excitation of the cell causes the proteins to alter their shape so that the central pore opens, and these channels are called **voltage sensitive**. The voltage-sensitive sodium channels open more quickly than the potassium channels so that at first sodium ions, which are more concentrated outside the neuron than inside, tend to pass into the cell. This continues until the electrical gradient across the membrane balances the concentration gradient for sodium ions, which occurs when the inside of the cell is about 40 mV more positive than the extracellular fluid. This is why the intracellular voltage reaches +40 mV at the peak of a spike. The spike is a brief event because the sodium channels do not remain open for very long, and voltage-sensitive potassium channels open at about the same time as the sodium channels close, so that potassium now leaves the neuron and the membrane repolarises towards resting potential.

The number of ions that flow during a spike is only a small proportion of the total number of ions in the cytoplasm and extracellular fluid. Nevertheless, the neuron needs to keep its batteries topped up by maintaining a difference in concentrations of sodium and potassium between its inside and outside. It does this by means of pumps, proteins in the membrane that consume metabolic energy to transport ions against concentration gradients.

2.3 Synapses

Neurons communicate with each other and other cells at specialised junctions called **synapses**, a term introduced by Charles Sherrington in 1897. Sherrington was convinced, from his physiological and anatomical studies of reflex functions of the spinal cord, that the transfer of signals between cells is a different type of process from the transfer of signals within a cell. It was not until the electron microscope was developed, 50 years after this, that the structure of synapses could be examined in detail. Synapses are tiny, and also very diverse. Sometimes a functional connection between two neurons is composed of just a single synaptic structure but more commonly there are several. Sometimes thousands of discrete anatomical contacts make up a connection between two neurons. A complex neuron such as a vertebrate motor neuron typically receives several tens of thousands of individual synapses from a variety of sources.

The neuron that is passing a signal on is referred to as **presynaptic**, and a neuron that is receiving a signal is referred to as **postsynaptic**. Transmission

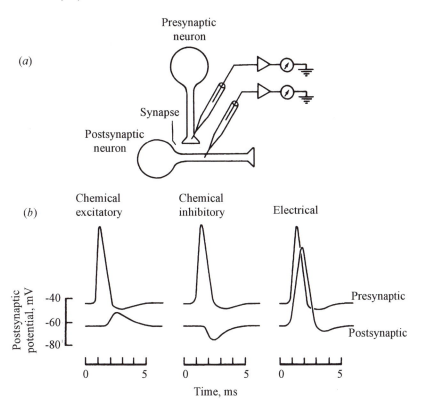

Figure 2.7 Operation of the three main types of synapse between nerve cells. (*a*) Schematic arrangement for recording the transfer of signals across a synapse by inserting electrodes close to the presynaptic and postsynaptic sites. (*b*) Recordings of presynaptic and postsynaptic potentials at the three types of synapse.

across a synapse can be studied by using two microelectrodes, one to record presynaptic potential and the other to record postsynaptic potential (Fig. 2.7). A significant majority of synapses are **chemical synapses** at which signals are transferred by way of chemicals such as acetylcholine or some amino acids that act as **neurotransmitters**.

The postsynaptic signal is called the **postsynaptic potential**, usually abbreviated to PSP. There is a delay, usually of about a millisecond, between the spike and the PSP. Generally, a spike in a presynaptic neuron, about 100 mV in amplitude, causes a PSP that is much smaller, a few millivolts in amplitude, although a few synapses are specialised to act as relays where the PSP is strong enough to trigger a spike in the postsynaptic neuron. At some synapses, the PSP excites the postsynaptic neuron by depolarising it and these PSPs are called **excitatory postsynaptic potentials** (EPSPs). At other synapses, the PSP hyperpolarises and inhibits the postsynaptic neuron and these PSPs are **inhibitory postsynaptic potentials** (IPSPs).

When a neuron is excited, it is more likely to release neurotransmitter and so pass signals to its postsynaptic targets. This is because the amount of neurotransmitter that is released depends on the membrane potential at the presynaptic sites. The way that membrane potential is linked to neurotransmitter release is by calcium ions which enter the presynaptic terminal through voltage-sensitive channels. Like the voltage-sensitive sodium channels that cause the upswing of a spike in an axon, these calcium channels open in response to depolarising signals. When a spike invades a presynaptic terminal, it depolarises the membrane strongly for a short time, so there is a strong pulse of calcium entry into the terminal and a brief squirt of neurotransmitter is released. Most of the delay in transmission across a chemical synapse is due to the time it takes for the voltage-activated calcium channels to open.

Spikes are not necessary for neurotransmitter release, and the link between membrane potential and the rate of neurotransmitter release is used by many neurons to transmit information in a graded manner. Such neurons have been studied in various visual systems (see section 5.3) and the motor systems of arthropods (see section 8.8), and are called **nonspiking** neurons because they normally operate without producing spikes. Instead, small variations in membrane potential regulate the amount of neurotransmitter release, probably because the rate of calcium entry into the presynaptic terminals is directly controlled by voltage-sensitive calcium channels. Rather than being released in squirts, neurotransmitter tends to dribble from the presynaptic terminals of these neurons, and some of them sustain a steady leakage.

Neurotransmitter diffuses extremely rapidly across the cleft that separates the membranes of the presynaptic and postsynaptic neurons. Some of its molecules attach to **receptor** proteins on the postsynaptic membrane and cause the shape of the receptor to alter. Some receptors are parts of ion channels, and a frequent action of neurotransmitter is to cause this type of ion channel to open. As with voltage-sensitive channels, these chemical-sensitive channels allow particular types of ions to pass through when they are open, and the direction and strength of flow are governed by the concentration and electrical gradients across the membrane. Many of these transmitter-activated channels allow sodium ions to pass through, and sodium will tend to enter the neuron, exciting it by causing an EPSP. Other transmitter-activated channels allow chloride ions to pass, and these ions tend to enter the neuron causing an IPSP. Potassium-conducting transmitter-activated channels also mediate IPSPs. Integration involves weighing up the balance of EPSPs and IPSPs within a cell and this determines how excited it is.

At a few synapses, electrical current can flow directly from one neuron to another. These are called **electrical synapses** (see Heitler, 1990, for a review). Under an electron microscope, an electrical synapse can be recognised as a region where the cell membranes of two neurons come close and touch. Protein channels called connexons link the two cells so that, when the synapse is active, the cytoplasm in the two cells is in direct contact, forming a conductive pathway for the electrical current. Sometimes electrical synapses conduct equally well in both directions, so that each neuron is both presynaptic and postsynaptic. A signal passes from one cell to another with negligible delay at an electrical synapse (Fig. 2.7b), which means that information passes across electrical synapses slightly more rapidly than across chemical synapses. Possibly associated with this, some of the best-known electrical synapses occur in pathways concerned with rapid escape responses (see Chapter 3). Electrical synapses can also help ensure that the neurons they connect are excited synchronously, which is useful for the co-ordination of some motor activities. This is also useful in some sensory systems such as the retina, in which electrical coupling helps to filter out real signals from extraneous noise. However, compared with chemical synapses, the scope for integration offered by electrical synapses is limited.

2.4 Integration of postsynaptic potentials

Integration involves weighing up the balance of EPSPs and IPSPs within a cell and the outcome determines how excited the neuron is at any given moment. In order to sort out all the PSPs that it receives, a neuron needs to have a way of combining them together. In general, the dendrites of a neuron function to combine PSPs as well as to receive them and this function is made possible by their **passive** cell membrane. This contrasts with the membrane of an axon, which is described as being **active** because it contains the voltage-sensitive channels responsible for generating spikes.

It is difficult to study the way in which PSPs travel in the dendrites of most neurons because of their complex branching structure. However, an axon or a muscle cell conducts potentials that are below the threshold for a spike in the same way as a long, unbranching dendrite would, and can be used to illustrate the properties of a signal as it is conducted along a length of passive membrane (Fig. 2.8a). These properties can be examined by using a pair of microelectrodes, one to inject pulses of current and the second to record voltage changes at different distances away from the point of injection.

Two important changes occur as a postsynaptic potential is conducted along a passive membrane. One is that its amplitude decreases as it travels away from its point of origin. Signal size is not directly proportional to the distance, but declines exponentially. This can be understood by considering the flow of electrical current along and across the cell membrane (Fig. 2.8b). When the cell membrane is passive, an axon or dendrite behaves electrically like an insulated cable and this type of current flow was studied by the engineers who first constructed long submarine cables to carry telegraph messages. The passive conducting properties of axons and dendrites are therefore often described as their **cable properties**.

The electrical model of a cell membrane in Fig. 2.8b consists of a network of electrical **resistors**, each of which offers a pathway for the flow of current. The current spread along the axon will be greater if the membrane resistance is high in relation to the internal resistance because less current will then flow out through the membrane. The cell membrane is indeed a relatively poor conductor of electricity, being composed chiefly of lipid, which is a good electrical insulator. Ion channels are the major route for current flow across the membrane and without them even less current would be siphoned away from the pathway along the length of the axon. The electrical resistance of the cytoplasm depends on the diameter of the axon, with wider axons being better conductors than narrow ones. Signals are therefore conducted passively for greater distances along wide axons and dendrites than along narrow ones.

The way in which a signal changes in size along the length of an axon or dendrite is expressed as the **space constant**, which is defined as the distance over which a signal decays to 37 per cent of its original amplitude

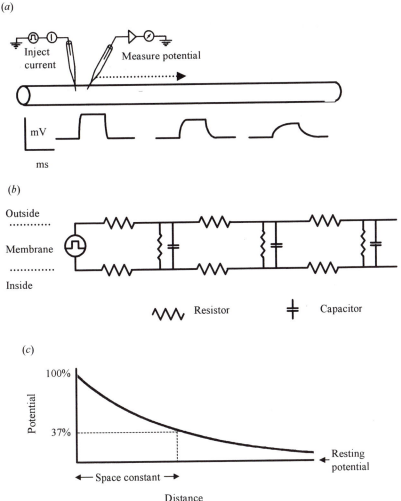

Figure 2.8 Cable properties of a length of axon or dendrite. (*a*) Two electrodes are inserted into the axon, one to inject a pulse of current and the second, which is moved successively further from the first, to record membrane potential. Drawings of voltage responses to a square current pulse are drawn for three locations. Note how the signal changes in size and shape as distance from its point of origin increases. (*b*) Electrical circuit for the membrane. (*c*) Graph to show how membrane potential declines with distance, including the definition of space constant for a cable.

(Fig. 2.8c). The space constant may be as great as 1 cm in a really wide axon, such the giant axon of a squid which has a diameter of 1 mm. It is difficult to measure the space constant in small neurons, but values from 0.2 to 1 mm have been estimated for typical mammalian dendrites. These values are large in relation to most neurons, which means that postsynaptic potentials will readily carry along the length of the average dendrite.

The second change that happens to a signal as it travels along a passive membrane is that its waveform becomes more rounded. This can be seen by comparing a square-shaped pulse of electrical current injected into an axon at one point with the responses at two locations further along (Fig. 2.8a). This change in shape is due to electrical **capacitance** of the membrane, a property that occurs because the relatively high-resistance membrane separates two low-resistance solutions having unlike charges (negative inside, positive outside). A capacitor stores electrical charges and a store, whether an electrical capacitor or a bucket of water, takes time to fill and empty. As the signal decreases in size with distance, it also takes longer to alter the charge on the membrane capacitor and this is why the waveform recorded in an axon or dendrite changes more slowly as the distance from the origin of the signal increases. A measure of the effect an axon or dendrite has on the time course of a signal due to its capacitance is given by its **time constant**, which is defined as the time taken for a constant current pulse to change the membrane potential to 37 per cent short of its final value.

When the cell membrane is responding passively, the potential change is graded in amplitude with the size of the electrical stimulus, in approximate accordance with Ohm's law. A number of these **graded potentials** can therefore sum, giving a resultant potential with an amplitude that is proportional to the current flow. The way in which the passive membrane of a neuron sums together different synaptic inputs is shown in Fig. 2.9.

In Fig. 2.9a a neuron is shown receiving one excitatory synapse, and the EPSP it produces is recorded by a microelectrode inserted into the cell body. A single EPSP depolarises the membrane, but not as far as the spike threshold for the postsynaptic neuron. If the presynaptic neuron spikes twice, the second EPSP sums with the first, and the postsynaptic neuron is more excited than by a single EPSP. A third EPSP is then able to carry the membrane potential past threshold. The ability to sum several postsynaptic potentials occurring in rapid succession depends on the time constant of the membrane. The capacitative properties of the postsynaptic

(a)

Excitatory synapse

Dendrite

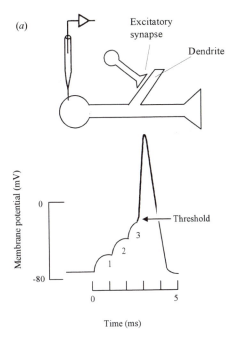

Membrane potential (mV)

0

Threshold

3

2

1

-80

0   5

Time (ms)

(b)

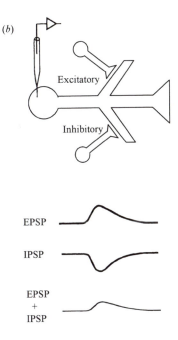

Excitatory

Inhibitory

EPSP

IPSP

EPSP
+
IPSP

(c)

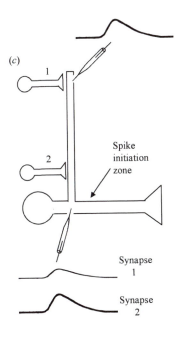

1

2

Spike
initiation
zone

Synapse
1

Synapse
2

Figure 2.9 Integration of postsynaptic potentials. In each section, an experiment to show a particular type of integration is shown schematically, together with the appearance of recordings on an oscilloscope screen. (*a*) Temporal summation of EPSPs. (*b*) Spatial summation of an EPSP with an IPSP. (*c*) Comparison of EPSPs at two locations along a long dendrite.

membrane slow the voltage changes so that an EPSP will decay more slowly than the synaptic current that gave rise to it. Consequently, a second EPSP, following soon after the first, will add its depolarisation to that remaining from the first EPSP. Several EPSPs in rapid succession are thus able to depolarise the membrane to threshold even though a single one is insufficient to do so. This is called **temporal summation.**

In Fig 2.9*b* a neuron is shown receiving an excitatory synapse on one dendrite and an inhibitory synapse on another dendrite some distance away. When both presynaptic neurons produce spikes simultaneously, an EPSP will be generated in the one postsynaptic dendrite at the same time as an IPSP is generated in the other dendrite. As these two potentials spread towards the recording electrode, they will meet and sum, producing a much reduced depolarisation in this case. Similarly, if two excitatory synapses are active at the same time, the postsynaptic neuron will be more strongly excited than if just one synapse were active. This integration of PSPs from different sites is called **spatial summation.** The ability to sum synaptic potentials occurring at different sites within the neuron depends on the space constant of the membrane. This is particularly important because the postsynaptic membrane itself is not usually capable of generating spikes; it is not electrically excitable. So the summed potential change produced by synaptic activity must spread passively to the nearest patch of membrane that is capable of generating action potentials. This is called the **spike-initiating zone**, and it is usually at some little distance from the synapses. For neurons with long axons, it is located close to the origin of the axon.

In this way, the sites of synaptic action and the sites of spike generation are restricted to different parts of the neuron, and are all linked together by the passive spread of graded potentials. This arrangement offers great scope for varying the integrative properties of neurons by varying the space constant, the dendritic geometry and the sites of synaptic input. For instance, the closer a given synaptic input is to the spike-initiating zone, the more influence it will generally have on the output, because its postsynaptic potentials will have decayed less than those of other inputs further away by the time they reach the spike-initiating zone. This is illustrated in Fig. 2.9 *c*, where two excitatory synapses are shown at different distances from the base of a long thin dendrite. An EPSP from the synapse towards the base of the dendrite will have decayed far less than one from the more distant synapse when recorded close to the spike-initiating zone. Similarly, an inhibitory synapse close to the spike initiating zone can effectively veto excitation that arises further out on the dendritic tree.

2.5 Comparison of spikes and graded potentials

Graded potentials and spikes are the universal language of neurons in all animals that have been studied. Nerve cells carry out their varied functions with just these two kinds of membrane potential. Spikes are clearly suitable for conveying information over long distances within the nervous system. Because spikes are relatively constant in amplitude, variation in signal strength is expressed in the frequency of spikes, with stronger signals being represented by higher numbers of spikes in a given time than weaker signals. The refractory period restricts the frequency of spikes that can be produced to a few hundred, or at most a thousand, per second. Graded potentials are used for short-distance communication and play an essential part at synapses and at sensory endings (see Chapter 4). Variation in signal strength is coded directly in the amplitude of membrane potential: the stronger a signal, the greater the change in membrane potential.

In electrical terms, graded potentials are analogue signals, whereas spikes are digital signals. In a nervous system, information is constantly transformed between these two types. Thus, chemical synapses transform digital signals into analogue signals by converting a burst of spikes arriving at the presynaptic terminal into a series of postsynaptic potentials. These graded variations in membrane potential can simply add together and so allow the neuron to perform its fundamental task of integration. In turn, analogue signals are transformed into digital signals at the spike-initiating zone of a neuron, where the frequency of spike generation depends on the rate of arrival of EPSPs. As larger and more numerous EPSPs spread to the spike-initiating zone, depolarisation of the membrane will occur more rapidly. So the threshold will be reached sooner and the time interval between successive spikes will be less, which means that the spike frequency will be higher.

2.6 Additional mechanisms in integration

The processes of spatial and temporal summation of PSPs in passive dendrites underlie much of the integrative activity of neurons, but additional features can also play significant roles. First, some dendrites have voltage-sensitive channels that can boost the amplitude of their input signals. One place where this has been studied is in output neurons of the thalamus of the vertebrate brain, which have the job of relaying sensory inputs to particular processing areas in the cortex. Second, there is not necessarily a physical separation between input and output regions of a neuron. Sometimes dendrites make output synapses as well as receiving inputs, and in some amacrine cells, input and output synapses are intermingled. This produces **local circuits** in the nervous system, where information can flow through processes of a number of different neurons without travelling very far. In the operation of local circuits, we cannot necessarily consider a whole neuron as an integrative unit because individual dendrites can act independently, summing local inputs to regulate output from their own synapses. The operation of local circuits has been particularly studied in the vertebrate olfactory system and retina (see Box 5.1, p. 107), and in circuits that control insect legs (see Chapter 8). Finally, the way that inhibition opposes excitation is not always by simple arithmetic summation of EPSPs and IPSPs. Strategically placed inhibitory synapses can make it hard for excitatory synapses to generate large EPSPs because, when postsynaptic channels at the inhibitory synapse open, they provide pathways for current flow through the cell membrane which effectively short circuits it. This reduces both the amplitudes of PSPs and, by reducing the neuron's space constant, the distance that they can travel.

2.7 Conclusions

Neurons consist of the same components as other animal cells, but are specialised for processing information. This specialisation is evident morphologically in the fine processes, the dendrites and axon, by which neurons make contact with each other. Physiologically, the most important specialisation is the maintenance of a membrane potential, which varies in response to incoming signals.

The transmission of signals over long distances is accomplished by spikes, which are active changes in membrane potential and are conducted along an axon with a constant amplitude at a finite speed. Signal strength is coded as spike frequency, a digital process. Over short distances, neurons use graded potentials that vary in amplitude and duration. These graded potentials usually originate at postsynaptic sites. Their amplitude reduces and their waveform becomes smoothed with increasing distance from the point of origin. Signal strength is coded as the amplitude of change in membrane potential, an analogue process. These graded potentials can be summed together: like variations in potential reinforce each other, and unlike variations tend to cancel each other out. This simple electrical property is the basis of most of the integrative processing that occurs in neurons.

Thus, the nervous system uses analogue processes to combine information and digital processes to transmit it over long distances. These are basically simple processes but a great variety of operation is achieved by adjustment of the physical properties of each neuron and its connections with other neurons. Physical properties such as membrane resistance, which affect the space and time constants, determine how far signals that originate at separate synapses are able to travel and combine together. The input and output connections of each neuron, reflected in its specific branching pattern, control the flow of information and create functional circuits in the nervous system. By combining the basic components and processes of neurons together in different ways, evolution has generated an extremely efficient system for processing information, one that is well able to control an animal's behaviour.

Further reading

Delcomyn, F. (1998). *Foundations of Neurobiology.* New York: Freeman. A comprehensive survey of basic neurobiology, covering a wide range of examples.

Nicholls, J.G., Martin, A.R & Wallace, B.G. (1992). *From Neuron to Brain: a Cellular and Molecular Approach to the Function of the Nervous System*, 3rd edn. Sunderland, MA: Sinauer. An established text which is widely used for its clear and thorough account of how nerve cells work.

Reichert, H. (1992). *Introduction to Neurobiology.* Stuttgart: Thieme Verlag; New York: Oxford University Press. A clear and concise account that introduces the reader to a range of topics.

3 Giant neurons and escape behaviour

3.1 Introduction

When an animal is suddenly attacked by a predator, it must respond with great urgency if it is to escape. The neuronal circuits that initiate such an escape response must be both straightforward and reliable in order to fulfil their biological function. A staightforward circuit is essential to ensure speed in initiating the escape, and a reliable circuit is needed not only to make sure the response occurs when required but also to avoid false alarms. These qualities of simplicity and reliability, which are of great survival value to the animal, are also of service to the neuroethologist exploring the role that nerve cells play in behaviour. Consequently, several of these startle responses have been studied in detail and they provide valuable insight into the flow of information through the nervous system from sensory inputs to muscular output.

Furthermore, these neuronal circuits often involve neurons that are exceptionally large and, because of this, are called **giant neurons**. The function of giant neurons is to conduct spikes rapidly along the body, but their size also makes them readily accessible to study with microelectrodes. The giant neurons therefore offer a major opportunity to investigate the role of individual nerve cells in behaviour.

Two main functions must be carried out by the neuronal circuit that initiates any behaviour pattern, including escape. First of all, a decision to initiate an activity must be made at some point in the circuit. This is usually done on the basis of incoming sensory information, which will often be filtered by the sensory receptors and neurons closely associated with them to extract particular stimulus features. This must occur rapidly because a startled animal has only a few milliseconds left to it in which to initiate escape action. The relevant processing must be fed to the decision point with minimal delay. Once the decision to initiate an escape has been made, the second function is an executive one: the circuit must include connections that lead away from the decision point to excite those neurons which are involved in the escape movement and to inhibit other neurons involved in incompatible movements.

Early work in this area gave rise to the idea that there may be a special class of high-order interneurons that carry out these two functions. The term **command neuron** was applied to such interneurons by Wiersma & Ikeda (1964), who found that electrical stimulation of single interneurons elicited co-ordinated movements of the abdominal ventilatory appendages of crayfish (swimmerets), even when the system was deprived of sensory feedback. Subsequently, a variety of instances was analysed in which apparently normal behaviour could be elicited by stimulation of a single neuron, and the term command neuron came into general use. Following controversy as to exactly what a command neuron might be, Kupferman & Weiss (1978) suggested that a command neuron should be defined as a neuron that is both necessary and sufficient for the initiation of a particular behaviour pattern. In this chapter, three cases of startle behaviour in which identified interneurons have been studied in detail are examined. Each case shows that this definition of a command neuron cannot be strictly applied to any of the neurons examined. This gives rise to a situation that will be entirely familiar to ethologists: one must either be content to use the term 'command neuron' loosely, or abandon its use altogether.

The three examples of startle behaviour considered here are the tail flip of crayfish, the fast start of teleost fish, and escape running by cockroaches. These have a number of features in common, including the fact that giant interneurons play a key role in initiating the behaviour in each case. The crayfish tail flip is probably understood, in terms of its neuronal circuitry, more completely than any other behaviour pattern of comparable complexity. Part of the reason for this relatively complete understanding is that recording from identified neurons has gone hand in hand with an increasingly exact study of the responses of intact animals, using film and video techniques. This illustrates well how ethological and neurophysiological methods of analysis can mutually reinforce one another during the intensive study of a single system.

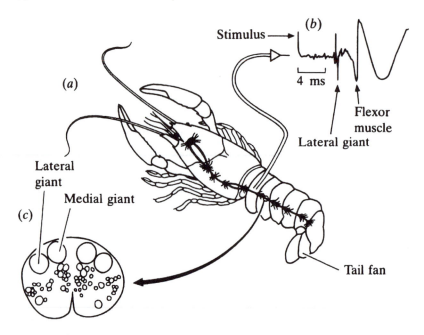

Figure 3.1 Giant interneurons involved in crayfish (*Procambarus*) startle behaviour. (*a*) Crayfish showing the location of the central nervous system (in solid black), a chain of ganglia. Also shown are electrodes implanted to record neuronal activity in the freely moving animal. (*b*) Activity recorded by these electrodes during a startle response: a tap on the abdomen (stimulus) is followed by a spike in a lateral giant and a potential in the abdominal flexor muscles. (*c*) Transverse section of the connectives between two abdominal ganglia, showing the locations of the lateral and medial giants. (*a* modified after Schramek, 1970; *b* redrawn after Krasne & Wine, 1975; *c* redrawn after Krasne & Wine, 1977.)

3.2 Giant neurons and the crayfish tail flip

Crayfish escape from the strike of a predator by means of a rapidly executed **tail flip**, produced by flexing and re-extending the whole abdomen. The abdomen is able to act as an effective locomotory organ because the last two (sixth and seventh) abdominal segments are modified to form the tail fan (Fig. 3.1*a*). A single flip of the tail fan is capable of moving the animal several centimetres through the water. The power for this movement is provided by the fast flexor muscles, which occupy much of the space within each abdominal segment. These are called fast muscles because they produce rapid twitch contractions, in contrast to a set of much smaller, slow muscles that produce graded postural movements of the abdomen. In addition, each abdominal segment contains fast and slow extensor muscles, and these are also much less substantial than the flexors.

The innervation of these muscles was first studied by Keis Wiersma (1947), who showed that the giant neurons are involved in the control of these muscles during a tail flip. About ten large motor neurons innervate the flexor muscles on each side of each abdominal segment. One of these is an exceptionally large motor neuron, called the **motor giant**, which sends an axon branch to every fast flexor muscle fibre. Another is an inhibitory motor neuron that also innervates every muscle fibre. The remaining motor neurons are simply known as fast flexor neurons, and each of these innervates only a localised group of fibres within the fast flexor muscles. During a tail flip, the motor giant and the fast flexor motor neurons are excited via two pairs of large interneurons, the lateral and medial giant interneurons (Fig. 3.1*b*, *c*). The control exerted by these two giant interneurons is so strong that a single spike in either a lateral or a medial giant interneuron is sufficient to trigger a tail flip.

Both of these giant interneurons extend along much of the length of the central nervous system but they differ considerably in structure. The medial giants have their cell bodies and dendrites in the brain, where they receive sensory input, and their axons extend down to the last abdominal ganglion. In contrast, the lateral giants are segmentally repeated structures, formed from separate cells linked end to end. Each segment contains a cell body, dendrites and a length of axon that abuts against the corresponding axon in the next segment. Where the axons abut, there is a segmental synapse between two successive lateral giants (see Fig. 3.3*a*). The lateral giants receive input only in the abdominal segments. At first, it was thought that both kinds of giant interneurons initiate the same behaviour pattern because a spike in either of them produces a rapid flexion of the abdomen. However, a more detailed series of studies, using a combination of neurophysiological and high-speed filming techniques, showed that the lateral and medial giants initiate different patterns of behaviour (Fig. 3.2).

Activation of the medial giants elicits contraction of the fast flexor muscles in all abdominal segments, and this produces a uniform curling of the abdomen that propels the animal straight backwards (Fig. 3.2*a*). Activation of the lateral giants elicits flexor contraction in the anterior segments of the abdomen but not in the posterior segments; the latter remain

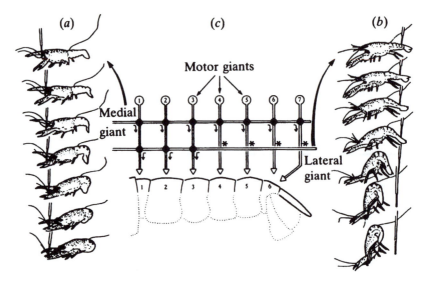

Figure 3.2 The different kinds of tail flip produced by the medial and lateral giant interneurons. The precise pattern of movement is correlated with activity in the giants by filming animals with electrodes implanted (as in *Figure 3.1a*). Tracings from these high-speed films show (*a*) a medial giant flip elicited by a tap on the head, and (*b*) a lateral giant flip elicited by a tap on the abdomen. (*c*) Diagram showing the pattern of synaptic connections between the two giants (horizontal lines) and the motor giant neurons (vertical lines) in the abdominal segments. Direct synaptic connections are represented by filled circles and the absence of synaptic connections is indicated by an asterisk. (From Wine, 1984.)

straight and so cause the thrust to be directed mainly downwards, thereby pitching the animal forwards (Fig. 3.2*b*). These two kinds of movements are well adapted to the different sorts of stimuli that excite the two types of giant interneuron. The lateral giants are triggered only by sudden mechanical stimuli that originate posteriorly, such as a sharp tap to the abdomen, and a lateral giant flip appropriately carries the animal in an anterior direction. Similarly, the medial giants are triggered only by stimuli applied to the head region and a medial giant flip carries the animal in a backward direction. In this way, each kind of tail flip removes the animal from the source of the stimulus.

The differences between the two kinds of tail flip can be explained by differences in the synaptic connections between the giant interneurons and the motor giant neurons in the abdominal segments (Fig. 3.2*c*). At the

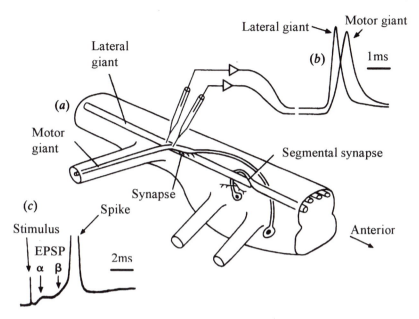

Figure 3.3 Electrical synapses between neurons involved in crayfish startle behaviour. (*a*) Drawing of an abdominal ganglion to show the relative positions of the lateral giant and giant motor neurons, the synapse between them, and the arrangement for recording from each side of this synapse. The segmental synapse between successive lateral giants is also shown. (*b*) Simultaneous intracellular recordings from the lateral giant and motor giant neurons close to the synapse, demonstrating the negligible delay due to electrical transmission. (*c*) Intracellular recording from the lateral giant at a point close to the synapse. Following electrical stimulation of the appropriate sensory neurons, a compound EPSP and a spike are recorded in the lateral giant with a very small delay. The two components of the EPSP (α and β) are produced by separate pathways from the sensory neurons, as described in the text. (*a* and *b* modified after Furshpan & Potter, 1959; *c* redrawn after Krasne, 1969.)

point of synaptic contact, the motor giant branches over the surface of the interneuron's axon in a characteristic manner, which is readily recognised when the motor axons are filled with intracellular dye (Fig. 3.3*a*). In the anterior segments of the abdomen, both lateral and medial giants receive these synaptic branches from the motor giants, but in the posterior segments the synaptic branches to the lateral giant are clearly missing, whereas those to the medial giant are present. This distribution of synapses has been confirmed by testing for postsynaptic responses by recording with a microelectrode: responses to medial giant activity can be obtained in all abdominal segments, but responses to lateral giant activity are only obtained in the anterior segments. In the thorax, the situation is reversed. The medial giant makes no output connections to motor giants, but the lateral giant does connect in the more posterior segments (Heitler & Fraser, 1993). Contraction of flexor muscles in the posterior part of the thorax helps bend the body into the jackknife shape that propels it rapidly forwards. Hence, the consistent difference in the pattern of abdominal flexion is brought about by differences in the synaptic connections between the respective controlling interneurons and a shared motor output system.

Another behavioural difference, noticed early in the study of crayfish startle behaviour, is that sometimes only a single tail flip occurs, while at other times an apparently similar stimulus evokes a whole series of tail flips in rapid succession. The latter are produced by alternating flexions and extensions of the abdomen, repeated at a frequency of 10 to 20 Hz, and this behaviour is termed escape swimming. Studying animals with implanted electrodes reveals that escape swimming does not involve the giants, which are active only before the first tail flip. The fast flexor muscles of the abdomen are used in swimming, and they are controlled by the fast flexor motor neurons but not by the motor giants and the giants. Escape swimming can be triggered on its own, without an initial giant-mediated flip, by stimuli that are weaker or have a slower onset compared with those which trigger the giant interneurons. Although a sharp stimulus can evoke a single giant-mediated flip without subsequent swimming, it is more usual for the giant-mediated flip to be followed by non-giant swimming.

The tail flips generated in escape swimming fall into three relatively stereotyped classes: linear flips, which resemble medial giant flips; pitching flips, which resemble lateral giant flips; and twisting flips, which tend to rotate the animal about its longitudinal axis as well as having a backward component. During swimming, these classes of tail flips are arranged in sequences that result in the animal being propelled backwards away from the original stimulus. Following a medial giant response, swimming involves only linear flips, which simply continue the backward movement away from the threat at the head end. A lateral giant response is followed by one or two pitching flips, which turn the animal in a complete somersault so that it lands on its back with its head facing the stimulus. Then two or three twisting flips turn the animal dorsal side up again, and finally a series of linear flips carry it backwards away from the stimulus (Cooke & Macmillan, 1985).

In addition, the swimming system is able to act upon directional information in the stimulus that is ignored by the giant system. The lateral giants, for example, always generate a bilaterally symmetrical response that carries the animal straight forward, regardless of whether the stimulus comes directly from behind or from one side. However, the path followed by subsequent swimming has a lateral component that steers the animal away from a stimulus delivered to one side of the abdomen (Reichert & Wine, 1983). All these results show that escape swimming is well adapted to exploit the initial advantage gained by a giant-mediated tail flip. The following account focuses on the lateral giant response and its relation to subsequent swimming because this system has been studied in most detail.

3.3 The lateral giant interneuron: input and output

Undoubtedly the most striking feature of the initial tail flip is its speed. Within 50 ms, abdominal flexion is completed and the animal has usually moved some distance through the water. The mean delay between the stimulus and the onset of flexor muscle potential is 6 ms (see Fig. 3.1b). This speed is required because the tail flip is probably a response to a predator that is extremely near to or touching a crayfish. Physiologically, the speed is partly achieved by extensive use of large neurons and of electrical synapses (see section 2.3). The segmentally repeated synapse between successive giant axons is electrical so that spikes are conducted across it with negligible delay and the chain of neurons acts effectively as a single axon. Each segmental lateral giant is also coupled via an electrical synapse with its contralateral partner, and with the motor giant neuron, which provides a pathway that conveys a spike from a lateral giant neuron to a fast flexor muscle in about 2 ms.

In the anterior segments of the abdomen, besides exciting the motor giants, the lateral giants also excite the other fast flexor motor neurons. A lateral giant makes synapses that have a mixed chemical and electrical nature with fast flexor motor neurons (Fraser & Heitler, 1991), but most excitation is conveyed by an indirect route, through an interposed neuron called the segmental giant because of the large size of its dendrite. There is a segmental giant on each side of each abdominal ganglion. Because of the

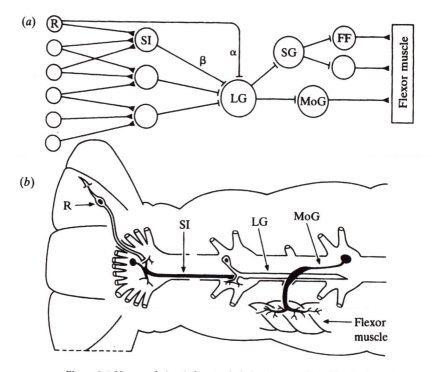

Figure 3.4 Neuronal circuit for startle behaviour mediated by the lateral giant interneuron. (*a*) Schematic representation of the excitatory pathway from the mechanoreceptors to the flexor muscles, showing chemical (—◄) and electrical (—|) synapses. Labelled circles represent: the receptors (R); sensory interneurons (SI); lateral giant (LG); segmental giant (SG); motor giant (MoG) and fast flexor motor neurons (FF). The sensory pathways that generate the two components of the compound EPSP (Fig. 3.3*c*) are labelled α and β. (*b*) Diagrammatic representation of the arrangement of the above components within the abdominal nervous system. The various components are not drawn to scale and only one segment of the lateral giant is shown; the medial giant and extensor motor neurons are not included. (Modified after Wine & Krasne, 1982).

negligible delay in the transmission of spikes from the lateral giant to fast flexor motor neurons across the two electrical synapses, the existence of the segmental giants was not suspected for many years. An unusual feature of the segmental giants is that they have a blind-ending axon in one of the lateral nerve roots, which might indicate that they were originally motor neurons. The output connections of a lateral giant are summarised in Fig. 3.4, which also shows the sensory input.

The lateral giants receive input exclusively from sensory neurons that are arranged in pairs, each pair being attached to a stout hair. These sensory structures occur on the dorsal surface of the abdomen, including the tail fan. The sensory neuron is a **bipolar neuron**, meaning that two processes extend out from its cell body, which lies at the base of a hair. One process, the dendrite, attaches to the inside of the hair, and is stretched when the hair is deflected, initiating a chain of events that leads to the production of an electrical signal. The second process is the axon, which carries signals in the form of spikes to the central nervous system. These sensory neurons are sensitive only to touch and to high-frequency water movements (about 80 Hz). They do not respond to low-frequency slosh of the water, which is detected by other hair receptors on the cuticle. Consequently, the hair receptors are well suited to detecting the shock wave in the water produced by the acceleration of a predator towards the crayfish, or even actual contact by the predator. This is a good example of **sensory filtering**: because these mechanoreceptors respond only to particular types of disturbance, the lateral giants, which they excite, will respond only to imminent danger, and not to water movements caused by waves or movements by the crayfish itself.

The main input pathway to the lateral giants runs from these receptors via sensory interneurons located in the segmental ganglia of the abdomen (Fig. 3.4). Most of these interneurons have relatively large dendrites and axons, from which synaptic potentials and spikes can be recorded readily with microelectrodes. The interneurons are an order of magnitude less numerous than the hair receptors, so there is considerable convergence in the input pathway. At the same time, there is some divergence, because each receptor axon branches to make contact with several interneurons. The receptors make chemical synapses onto the interneurons, which in turn make electrical synapses onto the lateral giant. A small proportion of the receptors make electrical synapses directly onto the lateral giant.

All the synapses on to the lateral giant are thus electrical but, because the postsynaptic neuron is so much larger than the presynaptic one, single spikes cannot generate enough current to depolarise the lateral giant above threshold. Instead, single spikes generate EPSPs which summate in the usual way (see section 2.4); hence, several spikes must arrive within a short time of each other at these synapses in order to trigger a spike in the lateral giant. Following stimulation of the hair receptors, a compound EPSP with two components can be recorded in the lateral giant (see Fig. 3.3c). The first component is due to receptors that synapse directly on the lateral giant, but this input alone is insufficient to drive the giant to threshold. The second component is due to input via the sensory interneurons. When the mechanical stimulus is strong enough, the second component sums with the first to carry the giant's membrane potential beyond threshold.

Thus, the lateral giant has a high threshold for spike initiation. Also, its membrane has a short time constant, which means that individual EPSPs die away quickly, so that many EPSPs must arrive within a short time of each other if they are to sum and trigger a spike. Together with the requirement of the mechanoreceptors for a high-frequency mechanical stimulus, this ensures that the lateral giant responds only to a sudden, abrupt stimulus. These characteristics are reinforced by the occurrence of habituation in tail flips mediated by a lateral giant: when a crayfish is tapped on the abdomen repeatedly, the probability of a tail flip in response diminishes rapidly. Stimulation at the rate of one tap per minute can diminish responsiveness to zero within ten minutes, and then many hours rest are needed for recovery. Physiological analysis has shown that one site at which habituation occurs is in the input pathway. The strengths of the chemical synapses between the receptor axons and sensory interneurons diminish (there is no change in the properties of the electrical synapses). However, habituation of the reflex is not simply due to an automatic reduction in the strengths of these synapses every time the sensory neuron is activated. Experiments by Krasne and Teshiba (1995) indicate that interneurons that originate in the brain exert a powerful influence over habituation, and ensure that habituation only occurs if stimulation of the sensory neurons is sufficiently intense to cause a tail flip.

3.4 The decision to initiate startle behaviour

It is clear that a lateral giant plays a crucial role in the circuit that generates a tail flip (Fig. 3.4). In fact, it is the place in the circuit where the decision is made to initiate this movement. Information from the hair receptors converges on the lateral giant and is integrated in its dendrites. The decision to initiate a tail flip is thus made when the compound EPSP reaches threshold and triggers a spike that is conducted along the axons of the lateral giants. Filtering in the sensory pathway ensures that a spike is only generated in response to appropriate stimuli. Once a spike is initiated in the lateral giant, the excitatory output to the motor giant produces contraction of all the fast flexor muscles in the most direct manner possible, and this is backed up by the slightly less direct route to the fast flexor motor neurons.

It can be shown that the lateral giant is essential to the production of a tail flip by hyperpolarising the cell so that it cannot produce spikes. When a lateral giant is prevented from firing in this way, stimuli that would normally cause a tail flip no longer do so (Olson & Krasne, 1981). For a given stimulus intensity, the amount of hyperpolarisation needed to abolish the tail flip is exactly that needed to prevent the lateral giant from firing. In the hyperpolarised cells, the size of the compound EPSP increases gradually as a function of the intensity of the stimulus delivered to the receptors, up to and beyond normal threshold level. This confirms that the lateral giant is the true point of decision in the circuit. If the decision were being made earlier, so that the lateral giant acted merely as a relay, one would expect to see an abrupt increase in EPSP size near threshold, due to firing of an earlier decision-making neuron. This effect is never seen.

The rapid flexion of the abdomen in a tail flip is a direct consequence of activity in the lateral giants, and it would be reasonable to assume that the same would be true of the subsequent re-extension, which follows within 100 ms of the original stimulus. However, this is not the case. Re-extension is a **reflex** initiated by particular mechanoreceptors stimulated by the flexion (Reichert, Wine & Hagiwara, 1981). That actual movement of the abdomen during flexion, and not just firing of the lateral giants, is necessary for re-extension is shown by recordings made from freely moving crayfish with electrodes implanted to record activity in the fast flexor and extensor muscles. If the abdomen is then restrained so that it cannot flex, or the motor nerves to the flexors are cut, no activity is recorded from the extensors following a stimulus that is adequate to excite the lateral giants.

No synaptic connections are apparently present by which the lateral giants could excite the extensor muscles after exciting the fast flexors during a tail flip. However, there are two appropriate classes of mechanoreceptors that do have appropriate connection with the fast extensor motor neurons. These are the hair receptors on the dorsal surface of the abdomen and muscle receptor organs, which each consist of a single sensory neuron with numerous branching dendrites that are attached to a slender receptor muscle. The receptor muscle lies parallel to a powerful muscle that extends the abdomen, and the muscle receptor organ, therefore, responds to movements that flex the abdomen. There is one phasic (fast, adapting response) and one tonic (slow, maintained response) muscle receptor organ on each side of each abdominal segment.

Both the tonic and the phasic muscle receptor organs make direct synaptic connections on to fast motor neurons of the extensor muscles. These are chemical, excitatory synapses, and each spike in the sensory neuron triggers a discrete EPSP in the motor neuron, with a synaptic delay of about 1 ms. Ordinarily, this reflex is used for making postural extensions of the abdomen in response to externally imposed loads, but it is equally suitable for triggering rapid re-extension following the internally produced flexion in a tail flip. The hair receptors also have a substantial excitatory input to the fast extensor motor neurons and trigger EPSPs with a longer delay, indicating a less direct pathway between sensory and motor neurons. Recording from crayfish with electrodes implanted in muscles shows that a waterborne stimulus to these hairs is able to excite brief activity in the extensor muscles of a resting crayfish. During a normal tail flip, then, the rapid water movement generated by abdominal flexion will excite the hair receptors, which will add their excitatory input to that of the muscle receptor organs and so activate the fast extensors of the abdomen.

The initial tail flip is normally followed by escape swimming, which consists of a series of tail flips that are not mediated by the giant axons. The lateral giants do not trigger escape swimming, any more than they do abdominal extension (Reichert & Wine, 1983). Nor is swimming triggered by sensory feedback from the first movements of flexion or extension. Instead, stimuli to the smooth hairs on the abdomen activate another pathway, that acts in parallel with, but more slowly than, the lateral giant pathway. Evidence for this is obtained by electrical stimulation of the lateral giants through electrodes implanted into resting, unrestrained crayfish. This stimulation elicits a rapid flexion followed by a re-extension, but it bypasses the pathways that are normally activated to trigger startle behaviour. In these circumstances, fewer than 1 per cent of all tail flips are followed by swimming. Such a result shows that neither activity in the lateral giants nor sensory feedback from the actual tail flip is sufficient to elicit escape swimming, so that this must be activated by an independent and parallel pathway that leads from the mechanoreceptive hairs.

Taps on the abdomen that are just below threshold for firing the lateral giants will often trigger escape swimming in well-rested crayfish. When this happens, the delay between the stimulus and the onset of the first flexor contraction is about 240 ms, slightly longer than the interval between the stimulus and the first flexion in a swimming bout that follows lateral giant spikes and a tail flip. This means that swimming is triggered through a much slower pathway than that responsible for triggering a tail flip, and that swimming is not simply inhibited during a tail flip. If inhibition was operating, one would expect swimming to begin with a shorter delay when it occurs without a preceding lateral giant response, but this is not the case. In fact the opposite is true: the average delay for swimming that is preceded by a giant-mediated tail flip is somewhat less than the delay for swimming that occurs without an initial tail flip (184 ms compared with 240 ms). This indicates that the lateral giants may facilitate the onset of swimming, although they do not trigger it. Hence, the smooth transition from a tail flip to swimming in normal escape behaviour is largely due to the activation of two different pathways that activate the two types of movement in sequence.

Escape swimming actually begins with a burst of activity in the abdominal extensor muscles, preceding the first flexor contraction by some 50 ms. This extensor activity is an event that is distinct from the reflex re-extension of the abdomen following a giant-mediated flip, but the two events often overlap when the delay to the onset of swimming is shorter than average. The extensors lead the flexors throughout swimming, with a nearly constant delay between extension and flexion over a wide range of swimming frequencies. As the frequency of abdomen movements declines during a swim, the delay between flexion and subsequent extension lengthens and the animal coasts with the abdomen in the flexed position; then it rapidly extends and flexes it again at the start of the next cycle. Evidently, extension of the abdomen during swimming is not produced by sensory reflexes, which must be inhibited. This and other evidence combine to suggest that the alternating extension and flexion during swimming is produced by a pattern generator consisting of networks of interneurons that work in a similar way to those described in Chapter 7 for other activities.

3.5 Executive functions of the lateral giant neuron

A lateral giant not only initiates a tail flip, but also extensively co-ordinates the sequence of events involved. This executive function is achieved by a massive and widely distributed array of inhibitory effects that follow a spike in the giant axon. Pathways lead away from the lateral giant to exert inhibition at almost every point in the neuronal circuit generating a tail flip. The IPSPs produced at these points in the circuit differ from one another, in the delay to their onset and in duration, in such a way as to ensure that each part of the response begins and ends at the right time.

The extensor motor system is the first place where inhibition is seen following a spike in a lateral giant axon. This inhibition is accomplished by parallel actions at three points in the motor pathway to the extensor muscles: the muscle receptor organs, the motor neurons, and the extensor muscle fibres themselves. The muscle receptor organ is inhibited by a special accessory cell associated with its sensory neuron; and the fast extensor muscles are inhibited by a motor neuron that causes IPSPs in the muscle fibres, called the extensor inhibitor motor neuron. The lateral giant excites these two kinds of inhibitory neuron, and also inhibits the excitatory motor neurons of the extensor muscles (Fig. 3.5a). These inhibitory actions begin within a few milliseconds of a lateral giant spike; in fact, the delay to the onset of the IPSPs in the extensor motor neurons is as short as that of the EPSPs in the fast flexor motor neurons (Fig. 3.5b). The duration of the IPSPs produced at all three points in the extensor pathway is relatively short, having an average value of 30 ms.

These arrangements clearly function to ensure the accurate timing of the reflex extension following giant-mediated flexion. The early onset of this inhibition prevents premature activation of the extensor reflex from interfering with the initial flexion. Inhibiting the extensor pathway at three separate locations makes quite certain that extension cannot occur while the inhibition lasts. The average duration of the IPSPs at the three locations is about the same as the average duration of the giant-mediated flexion, and so the extensor system is released from inhibition just as flexion is completed. The inhibitory action of the lateral giant thus co-ordinates flexion and re-extension effectively, even though the extensor system does not receive any additional excitation from the lateral giant.

Co-ordination of the response is continued by inhibition of the flexor motor system while re-extension takes place. Two inhibitory neurons have been identified that are important in shutting down the flexor system: the inhibitory motor neuron which innervates every fast flexor muscle fibre; and the motor giant inhibitor, an interneuron which prevents the motor

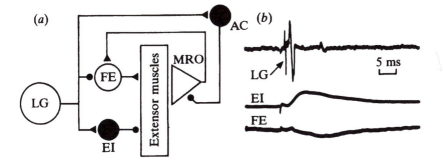

Figure 3.5 Inhibition of the abdominal extensor muscles by the lateral giant. (*a*) Neuronal circuit generating inhibition of the extensors, showing representative neurons: the lateral giant (LG); fast extensor motor neuron (FE); extensor inhibitor (EI); the muscle receptor organ (MRO); and its accessory cell (AC). Inhibitory neurons are shown in solid black; excitatory (—◀) and inhibitory (—●) connections are shown, but electrical and chemical synapses are not distinguished. (*b*) Typical recording used to build up the interpretation given in (*a*). The upper trace is an extracellular record from the motor nerve to the flexors, showing the lateral giant spike and subsequent compound spike from flexor motor neurons. The middle and lower traces are intracellular records showing, respectively, an EPSP in the extensor inhibitor and an IPSP in the fast extensor motor neuron. Note the short delay of the postsynaptic potentials after the lateral giant spike. (*b* from Wine, 1977; copyright Springer-Verlag.)

giant from firing. Both of these inhibitory neurons are excited indirectly by the lateral giant (Fig. 3.6*a*). The motor giant inhibitor is strongly excited by the fast flexor motor neurons and so brings about inhibition of the motor giant very soon after the latter has fired. The flexor inhibitor is weakly excited by a variety of sources, including the fast flexor motor neurons and the corollary discharge interneurons, a class of interneurons which are excited by the segmental giants and with axons that run between segments. These sources of input bring about early depolarisation of the flexor inhibitor, but the threshold for spiking is reached only after the corollary discharge interneurons in other ganglia are recruited to provide additional input.

This roundabout pathway increases the delay, and consequently a spike is not triggered until about 15 ms after the spike in the lateral giant (Fig. 3.6*b*), by which time the motor giant is already inhibited. The actual flexion movement has only just begun at this time; but it must be remembered that it takes a few milliseconds for the muscles to develop tension and for the

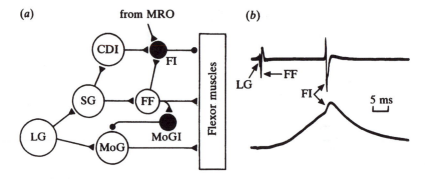

Figure 3.6 Delayed inhibition of the fast flexor muscles by the lateral giant.
(*a*) Neuronal circuit generating inhibition of the flexors, showing representative neurons as in Fig. 3.4 plus: the motor giant inhibitor (MoGI); corollary discharge interneuron (CDI); and flexor inhibitor (FI). Symbols have the same meaning as in Fig. 3.5, and chemical and electrical synapses are not distinguished. (*b*) Recording demonstrating the delayed activation of the flexor inhibitor. The upper trace is an extracellular record showing the delay between the spike in the lateral giant and the spike in the flexor inhibitor; the lower trace is a simultaneous intracellular record from the flexor inhibitor showing the compound EPSP gradually rising to threshold. (*b* from Wine & Mistick, 1977; reprinted by permission of Wiley-Liss, Inc., a subsidiary of John Wiley & Sons Inc.)

tension to overcome inertia in the skeleton, so that by the time actual movement begins the electrical events that triggered the movement have already been completed. Hence, the delayed inhibition from the lateral giant prevents any additional flexor activity and prepares the way for re-extension. In addition, the flexor inhibitor receives an excitatory, monosynaptic connection from the muscle receptor organ and this input becomes active as soon as the muscle receptor organ is released from inhibition. This input sums with that from the lateral giant, thereby prolonging activity of the flexor inhibitor for the duration of re-extension.

A further important place where inhibition acts is on the input side of the circuit, at the synapse between the hair receptors and the sensory interneurons. This inhibition acts both postsynaptically, on the dendrites of the interneurons, and also presynaptically, on the axon terminals of the receptor neurons. **Presynaptic inhibition** of this type is common in mechanosensory systems, and its action is to reduce or prevent the release of neurotransmitter. Both postsynaptic and presynaptic inhibition are delayed to about 15 ms after the lateral giant spike due to the indirect pathway that mediates them, including the corollary discharge and other interneurons. This inhibition has a long duration of about 50 ms. However, not all the hair receptors are inhibited in this way; those that provide input to the extensor motor neurons are not inhibited and they contribute to the re-extension reflex.

The presynaptic inhibition of the first input synapse plays an important part in co-ordinating the startle reflex because abdominal flexion is a dramatic movement that stimulates the same receptors that trigger the reflex. If this input were not inhibited, it could cause a perpetual cycle of repeated tail flips by positive feedback. As it is, the onset of inhibition coincides with the onset of movement of the abdomen and so prevents this feedback effect. The long duration of the inhibition makes sure that a second tail flip is not triggered before the first one is completed. The occurrence of presynaptic inhibition at the first synapse is well correlated with the fact that this synapse is the site of habituation of the response. If postsynaptic inhibition alone were present, then synaptic transmission could still habituate during the repetitive stimulation caused by abdominal flexions, and this could render the whole system unresponsive for several hours. This habituation does not occur because presynaptic inhibition prevents the presynaptic terminals from being fully activated.

3.6 Summary of pathways in crayfish startle behaviour

The startle response of crayfish is a simple behaviour, which is initiated with the least possible delay. Nevertheless, the neuronal circuit underlying this behaviour is quite sophisticated and involves complexities that could hardly be predicted from the behaviour itself. The main pathways involved in linking the response to an adequate stimulus are shown in Fig. 3.7. The initial tail flip is triggered by sensory information, which is fed with minimal processing on to the lateral giant, which acts as a command neuron. The lateral giant produces excitation of the flexor muscles and, at the same time, short-lasting inhibition of the extensors. This is followed by delayed, long-lasting inhibition of the flexors and of sensory input to the lateral giant itself. Finally, excitation of the extensor muscles is produced by sensory feedback from the first flexion. Through the operation of these pathways, the first tail flip is completed (to full re-extension) by about 110 ms from the initial stimulus.

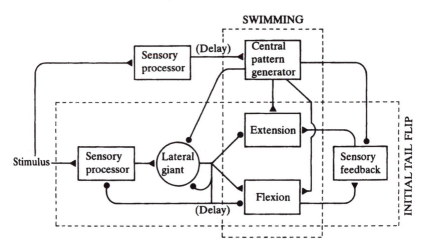

Figure 3.7 A flow diagram summarising the functional relations between the major components of the startle behaviour in crayfish. The usual symbols are employed to represent excitatory (—◄) and inhibitory (—●) relations, but here the labelled boxes do not represent individual neurons, except in the case of the lateral giant. The components of the initial, giant-mediated tail flip are enclosed in the horizontal, dotted rectangle, and those of non-giant swimming are enclosed in the vertical, dotted rectangle. The flexor component has some vertical elements that are not common to both systems, whereas the three separate sensory components may well have some elements in common. (Modified after Wine, 1984.)

Sensory information that is adequate to trigger the giant-mediated tail flip also triggers escape swimming by an independent pathway that involves more elaborate sensory processing. This elaborate processing enables the swimming system to take account of directional information in the stimulus that is ignored by the giant-mediated first tail flip and also introduces a considerable delay. The delayed excitation triggers a bout of swimming, in which the extensors lead the flexors in each cycle. The delay in triggering swimming is such that the first movement of escape swimming, an extension, overlaps with or immediately follows the re-extension of the first tail flip. When the crayfish swims, inhibition prevents both activation of the extensors, through sensory feedback from their receptor organs, and activation of the lateral giant system.

As the startle behaviour of crayfish continues to be studied, many additional complexities are coming to light; these include additional circuit elements such as non-spiking interneurons that control abdominal muscles and circuits that control walking legs. Finally, it should be noted that the startle response is not automatic, although it will override and inhibit any conflicting, competing activity once it has been triggered. The readiness with which the startle response can be triggered is modulated by a range of factors, including strong control of elements in the circuit by neurons descending from the brain that can make it almost impossible to elicit a startle response in some circumstances (Krasne and Teshiba, 1995).

3.7 Mauthner neurons and the teleost fast start

When a sharp tap is delivered to the side of an aquarium, the fish inside exhibit a characteristic startle response consisting of a brisk swivelling movement that displaces the fish sideways by a small amount. In natural circumstances, this is an effective escape movement that enables the animal to dodge the strike of a predator. The key neuron in this startle response is called the **Mauthner neuron**; there is one of these neurons on each side of the brain of most species of fish and of amphibians. Most studies of these neurons have been made in goldfish and zebra fish. Because of the exceptionally large size of Mauthner neurons, it has been possible to study them in both dissected preparations and intact animals; these studies provide one of the few cases in vertebrates in which a clear causal relationship has been established between activity in a particular neuron and performance of a specific behaviour pattern. The way the Mauthner neuron operates shows a number of instructive parallels with the crayfish lateral giant neuron.

In teleost fish, the startle movement initiated by a Mauthner neuron is known as a fast start, and consists of a stereotyped sequence of movements that occur in three stages. In the initial stage, the trunk muscles contract all along one side of the body so that the animal assumes a C-like shape with the head and tail bent to the same side. A number of other actions are also initiated, such as closing the mouth, drawing in the eyes and extending the fins. During the second stage, muscle contractions proceed down the other side of the body so that the tail straightens. These first two stages result in a sudden acceleration that propels the animal in a direction that is deter-mined by the extent of the body bend in the first stage and displaces the fish by about one body length. The third stage consists of normal swimming movements, or sometimes coasting, which carries the fish further away.

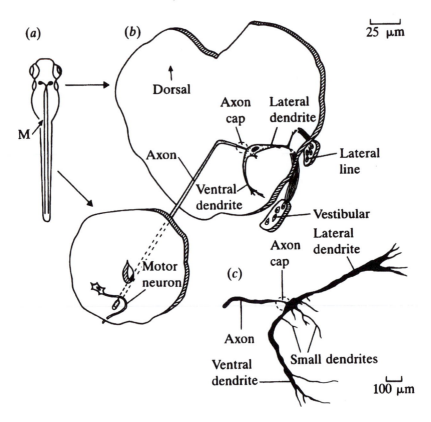

Figure 3.8 The Mauthner neuron in teleost fish. (*a*) The general location of the bilateral pair of Mauthner neurons (M) shown schematically for a larval zebra fish (*Brachydanio*). (*b*) The right Mauthner neuron is shown in a thick, transverse section of the hindbrain at the level of the eighth cranial nerve, together with the inputs from the lateral line and acoustico-vestibular systems. The output of the Mauthner axon to the spinal motor neurons is shown in a thick transverse section of the spinal cord. (*c*) The Mauthner neuron of an adult goldfish (*Carausius*), reconstructed from transverse sections of a cell injected with intracellular dye. (*a* modified after Prugh, Kimmel & Metcalfe, 1982; *b* modified after Kimmel & Eaton, 1976; *c* redrawn after Zottoli, 1978.)

The large cell body and two primary dendrites of a Mauthner neuron are located in the hind brain (Fig. 3.8). The axon crosses to the opposite side of the brain before descending the nerve cord to contact motor neurons of trunk muscles. This basic anatomy was described by Bartelmez (1915), who first suggested that it was involved in startle behaviour, although for several decades after that it was thought that the Mauthner neuron is involved in normal swimming movements. The role of the neuron in startle behaviour was firmly established when recordings from the neuron were correlated with movements in freely moving animals – spikes in a Mauthner neuron can be identified unambiguously in extracellular recordings because their waveform provides a characteristic signature. When a spike is recorded from a Mauthner neuron in response to a sudden sound, it is always imme- diately followed by a large muscle potential in the trunk muscles on the side of the body opposite to the Mauthner neuron cell body, and the fish per- forms a fast start (Fig. 3.9a).

The precise timing of the Mauthner spike in relation to the stages of the fast start is clarified by using an implanted electrode in conjunction with high-speed filming of the response (Fig. 3.9b). This method shows that the average delay from a stimulus to a Mauthner spike is about 6 ms. There is a further delay of about 2 ms until the muscle potential starts and then a delay of 6–10 ms until the muscle develops sufficient tension for actual movement to begin. The speed of this startle response is comparable to that of the crayfish tail flip, which is not surprising as both are natural responses for evading the strike of a predator. Another parallel with the crayfish startle system is that only a single spike occurs in the Mauthner neuron, and this precedes the initial C-like bend. Mauthner spikes do not accompany the second stage or subsequent swimming, which must therefore be due to the activity of other interneurons acting in parallel with the Mauthner neuron.

The Mauthner neuron on the side closest to the stimulus is the only one that produces a spike and, because this excites the contralateral muscula- ture, the initial C-like turn is made to the side away from the stimulus. The size of this initial turn is not constant but varies with the angle of the threat- ening stimulus with respect to the fish. This is shown by dropping a metal ball on to the water near the goldfish and analysing high-speed films of the resulting startle response. In this experiment, the angle through which the fish has turned by the end of stage one is inversely proportional to the direction of the impact of the ball (Eaton & Emberley, 1991). When the startle response is filmed in goldfish with electrodes implanted in the trunk musculature (as in Fig. 3.9 a), the results show that progressively larger turns in stage one are correlated with progressively longer and more complex muscle potentials. The fish is evidently controlling how far it turns by varying the activity of motor neurons supplying the trunk musculature.

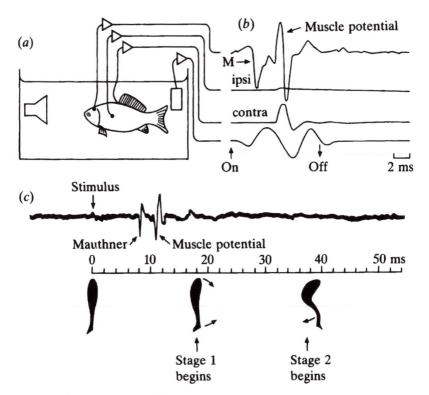

Figure 3.9 Activity of the Mauthner neuron during startle behaviour. (*a*) A goldfish in an aquarium with electrodes implanted for recording from one Mauthner neuron and from both left and right trunk muscles in a freely moving animal. The stimulus is delivered by a loudspeaker (at left) and monitored by a hydrophone (at right). (*b*) Typical recording of a startle response to a sound stimulus (bottom trace, On/Off). The brain electrode (top trace) picks up the Mauthner spike (M) and also the prominent potential that spreads from the trunk musculature. The two myogram electrodes (middle traces) show that this muscle potential originates from the contralateral muscles (contra) and not the ipsilateral muscles (ipsi). (*c*) Simultaneous recording with electrodes implanted in the brain and high-speed filming in dorsal view (representative silhouettes at bottom) serves to establish the precise timing of the Mauthner spike in startle behaviour. (*a* and *b* modified after Zottoli, 1977; *c* modified after Eaton, Lavender & Wieland, 1981.)

However, direct electrical stimulation of the Mauthner neuron consistently results in a short and simple muscle potential. Hence, the size of the C-like turn must be controlled by other interneurons that are active in parallel with the Mauthner neuron (Eaton, Di Domenico & Nissanov, 1991).

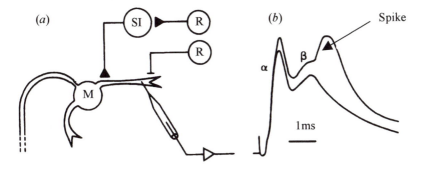

Figure 3.10 Sensory input to the Mauthner neuron. (*a*) Schematic representation of input to the lateral dendrite from the eighth cranial nerve, showing receptors with direct electrical (—|) synapses onto the Mauthner neuron (M) and those with indirect chemical synapses (—◀) by way of sensory interneurons (SI). (*b*) Intracellular recording from the distal end of the lateral dendrite, as indicated in (*a*), showing two superimposed records. Following electrical stimulation of the receptor axons, a compound EPSP is recorded, with an early component (α) due to the electrical synapse and a later component (β) due to the chemical synapses. In one of the records, the latter triggers a spike, which is small in size because it has spread passively from the spike-initiating zone to the recording site. (*b* modified after Diamond, 1968.)

3.8 Excitation and inhibition in the Mauthner neuron

The Mauthner neuron is excited by a wide range of sensory modalities; as a result, a large number of sensory neurons converge on this neuron, either directly or indirectly, and this is reflected in the great size of the two main dendrites (Fig. 3.8*c* and Fig. 3.10*a*). The cell body and the dendrites sum the synaptic potentials which are generated by inputs from the sensory neurons, and this summed signal is conducted passively to the initial part of the axon. The membrane of the Mauthner neuron has a large space constant, up to 5mm, which enables the dendrites to conduct passive signals several hundred microns from the more distant synapses towards the axon. The initial part of the axon plays a crucial part in the integrative mechanism of the Mauthner neuron because it is where spikes are initiated when the input delivered from the dendrites exceeds threshold – the dendrites and cell body are incapable of supporting spikes.

The input that has been studied in most detail is that coming from the auditory system. The underwater sounds that accompany the onrush of an attacking predator can be detected by the sensory receptors in the inner ear of the threatened fish. Using a loudspeaker (see Fig. 3.9a) or a high-pressure jet of water as a stimulus in experiments shows that acoustic stimuli are certainly capable of initiating fast starts. Visual stimuli and input to the lateral line may augment the auditory stimulus but do not appear to be capable of triggering a spike in the Mauthner neuron on their own. For example, in the above experiment in which a ball was dropped on the water, the goldfish could see the ball coming but a fast start was not initiated until several milliseconds after the ball hit the water. This suggests that the triggering stimulus for the Mauthner neuron was the acoustic one from the impact of the ball on the water.

The sensory neurons from the inner ear provide the major input to the lateral dendrite, where they make direct (monosynaptic) contact in the form of large club-shaped endings. When postsynaptic potentials are recorded by a microelectrode in the lateral dendrite, they clearly contain two components: the α component, which is an initial, sharply rising EPSP after a short latency of about 0.1 ms; and the β component, which is a more gradually rising EPSP with a longer latency, of about 1 ms (Fig.3. 10b). The latencies of these two events suggest that the first is mediated by electrical synapses, and the second by chemical synapses, possibly from sensory interneurons. The α component can be elicited by relatively weak shocks to the eighth cranial nerve, which excite the largest sensory axons but fail to excite the smaller sensory axons that are responsible for the chemical EPSP. Electron microscopy of the distal part of the lateral dendrites confirms that these axons synapse with the Mauthner neuron by way of **tight junctions** between the membranes of the two cells, which are characteristic of electrical synapses.

As stimulus strength is increased further, both the α and the β components grow in size, until their EPSPs are sufficiently large to trigger a spike. In the recording shown in Fig. 3.10b, the spike appears small and rounded, and the chemical EPSP that triggers it appears smaller than the preceding electrical EPSP. This is because the site of the recording was towards the end of the lateral dendrite, in the region where it receives the electrical synapses; both the spike and the chemical EPSP, which is caused by synapses closer to the cell body, were conducted passively along the cell back to the recording site.

Thus, the electrical EPSP has a priming role; in natural conditions, it promotes early depolarisation of the spike-initiation zone, so that a following, large, chemically mediated EPSP rapidly triggers a spike. As in the input circuit to the crayfish lateral giant, both electrical and chemical synapses play roles in conveying sensory information to the Mauthner neuron. The large diameter of the Mauthner neuron axon, together with extensive insulation with myelin, ensure that, once initiated, a spike travels rapidly down the spinal cord. In goldfish, the conduction velocity for this spike is 85 m/s, which enables the signals to travel the whole length of the spinal cord within 1 ms. Consequently, all the motor neurons to the trunk muscles on that side of the body are excited virtually simultaneously.

3.9 Outputs and executive functions of the Mauthner neuron

The Mauthner neuron has monosynaptic connections with many of the spinal motor neurons that it excites; a short collateral of the Mauthner axon makes contact with a special region of the motor axon several micrometres from the cell body (see Fig. 3.8*b*). A spike in the Mauthner axon produces an EPSP in the motor neuron with a synaptic delay of about 0.6 ms, so the transmission is probably chemical. The excitation of a substantial number of other motor neurons is relayed by way of a premotor interneuron and so involves a slightly greater delay. The motor axons travel a relatively short distance to the trunk muscles, where they have normal chemically transmitting synaptic endings on the muscle surface. This short and direct pathway is responsible for the short delay of about 2 ms between a Mauthner neuron spike and the onset of a spike in the trunk muscles (see Fig. 3.9*b*, *c*).

A spike in the Mauthner axon inhibits the spinal motor neurons contralateral to the axon at the same time as it excites those ipsilateral to the axon. The inhibition appears to be mediated by inhibitory interneurons that cross the midline and are activated by electrical synapses from the Mauthner axon (Fig. 3.11*a*). This crossed inhibitory circuit makes sure that, if only a short interval in time separates spikes in the left and right Mauthner neurons, only the earlier of the two Mauthner spikes is able to fire its motor neurons. If the two Mauthner neurons spike simultaneously, the crossed inhibition prevents any motor output at all.

Within the brain, branches from the axon of the Mauthner neuron excite a number of cranial relay neurons, which carry out some important

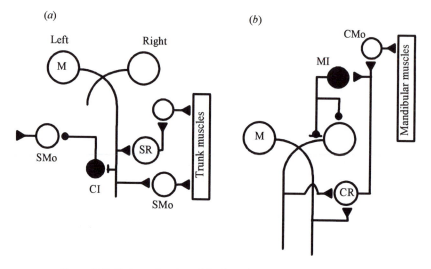

Figure 3.11 Output circuitry of the Mauthner neuron. (*a*) Neuronal circuit generating excitation and inhibition of spinal motor neurons, showing: the Mauthner neuron (M); spinal relay neurons (SR); crossed inhibitory interneuron (CI); and spinal motor neurons (SMo). Chemical excitatory (—◀), electrical excitatory (—|) and chemical inhibitory (—●) synapses are shown. (*b*) Neuronal circuit generating excitation of the cranial motor neurons and self-inhibition, showing the Mauthner neuron (M); cranial relay neuron (CR); cranial motor neurons (CMo); and the Mauthner inhibitor (MI). Synaptic symbols are as in (*a*), plus electrical inhibition of the axon cap (—●|).

functions in the startle response (Fig. 3.11 *b*). The cranial relay neurons excite motor neurons of muscles that close the jaw and draw in the eyes, which helps to streamline the fish for its escape. In addition, the cranial relay neurons monosynaptically excite a group of interneurons that inhibit both the Mauthner neurons.

These Mauthner inhibitors have cell bodies clustered around the ventral dendrite and they act to inhibit the Mauthner neuron in several ways. Some of them exert an unusual form of electrical inhibition: they have axons coiled tightly round the origin of the Mauthner axon, contributing to the axon cap (see Fig. 3.8*b*). When spikes are generated in these axons, they make the outside of the Mauthner neuron locally more positive, which has the same effect as driving the intracellular potential more negative. Spikes from these neurons thus exert a kind of brief stranglehold on the Mauthner neuron, right at the zone where its spikes are initiated. This electrical inhibition follows a Mauthner spike with a delay of about 1 ms and so prevents the Mauthner neuron from firing twice in response to a given stimulus. Furthermore, the cranial relay neurons are excited by branches of both left and right Mauthner axons, with the result that the contralateral Mauthner neuron also receives electrical inhibition with the same short delay after a spike in the ipsilateral axon.

The electrical inhibition of both Mauthner neurons is a brief event, lasting no more than about 2 ms, but it is followed by inhibition through chemical synapses that lasts much longer. The Mauthner inhibitors produce this inhibition through their axon terminals on the cell body of the Mauthner neuron, which make conventional chemical synapses. The chemical inhibition begins about 1 ms after the electrical inhibition, and lasts for about 45 ms, which takes it well into the second stage of the startle response. There is evidence that this direct inhibition is backed up by pre-synaptic inhibitory synapses made by the Mauthner inhibitors on to the synaptic terminals of the sensory neurons on the Mauthner neuron dendrites, and also by separate inhibitory circuits that are activated directly by the sensory neurons. The result of these inhibitory mechanisms is not only that the active Mauthner neuron and its contralateral partner are shut down as soon as a spike has been generated, but also that they are kept shut down long enough for the startle response to go to completion.

Although the involvement of Mauthner neurons in triggering startle responses in teleost fish is clear, this does not mean that this is the only function that these neurons serve. Sudden lunges by goldfish to capture prey on the water surface involve movements similar to those of a startle response, and follow a Mauthner neuron spike (Canfield & Rose, 1993). Besides integrating information about a source of potential danger, a Mauthner neuron must be able to collect information that triggers an attack at the appropriate time. The input and output circuitry of a Mauthner neuron is arranged so that it can initiate and co-ordinate a rapid, directed movement.

3.10 The startle reaction of a cockroach

Cockroaches are extremely difficult to catch; this is partly because they are exquisitely sensitive to very small air currents which trigger fast-running responses that are accurately directed away from the source of wind. In

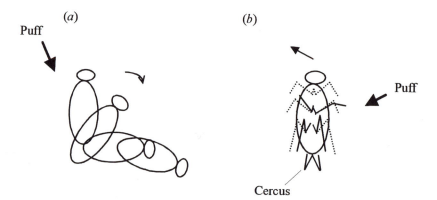

Figure 3.12 Startle response of the cockroach. (*a*) Video recording, made from above the cockroach, of the turning response to a wind puff delivered from the front left of the animal. The outlines of the body and head are traced from every second frame. (*b*) Leg movements during an escape turn caused by a puff of wind coming from the right and slightly to the front. The cockroach was held so that its body could not move, but its legs slipped over a lightly oiled glass plate. The initial positions of the legs are indicated with dotted lines; the final positions of the middle and hind legs are drawn as solid lines. The arrow above the cockroach indicates the direction in which the animal would have faced if its body had been free to rotate. (*a* redrawn after Comer & Dowd, 1993; *b* redrawn after Ritzmann, 1993.)

their natural habitat, which for the American cockroach *Periplaneta americana* is leaf litter on the forest floor, cockroaches have been shown to be able to evade the strikes of one of their predators, the toad, by detecting the movement of air caused by movement of the toad's tongue (Camhi & Tom, 1978). Cockroaches are particularly sensitive to sudden increases in the velocity of air currents, which is how they can distinguish an attack by a toad from meteorological wind (Plummer & Camhi, 1981).

Unlike the startle response in crayfish, cockroach startle behaviour is normally controlled by activity in several giants, and involves many spikes in these neurons rather than the single spike that marks the decision-making process in the crayfish lateral giant or Mauthner neuron. Undoubtedly associated with this, the cockroach startle response is initiated more slowly than that of the crayfish or fish, but is nevertheless a rapid behaviour, with the first movement occurring within 50 ms of the start of a wind stimulus. It is a rapid turn away from the direction of attack (Fig. 3.12*a*) that, unlike walking or running, involves the co-ordinated movement of all six legs at the same time (Fig. 3.12*b*). Each leg pushes or pulls, so that the cockroach swivels to face away from the direction from which the air current is coming. The cockroach then runs forward, using the usual tripod gait mode of locomotion in which the legs are moved in two sets of three.

The sense organs that excite the giants are called filiform sensilla; each is a slender, whip-like structure inserted in a socket on the ventral surface of a cercus, the sensory structure that projects from each side of the last abdominal segment. The sensillum articulates in its socket, and can move in one particular plane. It is attached to a single bipolar sensory cell that is excited when its sensillum is deflected in one direction. The axon of the sensory cell enters the last abdominal ganglion, where it terminates in a series of fine branches that are restricted to one side of that ganglion and which make chemical synapses on to dendrites of the giant interneurons. The filiform sensilla are arranged in 14 columns, most of which run the length of the cercus. Sensilla of a particular column all share the same direction for deflection, and so all respond best to air currents coming from the same direction. A cercus of an adult cockroach has about 20 segments, and each column is usually represented by one filiform sensillum on each segment. The arrangement for two adjacent segments is shown in Fig. 3.13*a*.

Seven pairs of interneurons that originate in the last abdominal ganglion and have axons that run in the ventral nerve cord have been labelled as giants. The axons range from 20 to 60 μm across, so they are considerably smaller than the giants of crayfish or Mauthner neurons, but are appreciably larger than other axons in the cockroach nerve cord. Anatomically, two separate groups of giants can be distinguished, one with axons located dorsally and the other with axons located ventrally in each connective nerve. The giants in the ventral group are larger than those in the dorsal group and trigger escape running. The structure of one ventral giant, number 1, is shown in Fig. 3.13*b*.

Carefully conducted experiments have revealed that each giant responds most vigorously to air currents from a particular direction (Kolton & Camhi, 1995). To determine the directional selectivity of a giant, spikes were recorded from its axon using a microelectrode that contained a stain so that the neuron could later be identified from its anatomy. Carefully controlled puffs of air, always with a peak velocity of 0.85 m/s, were directed at the

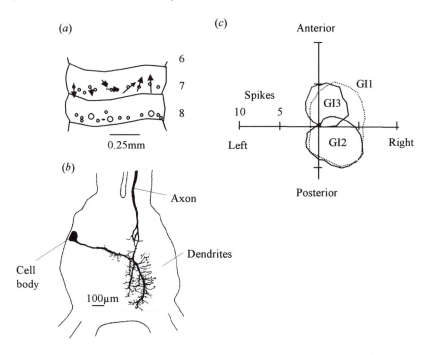

Figure 3.13 Giant interneurons and sensory analysis of wind-puff direction in the cockroach. (*a*) Best directions of deflection for exciting sensory neurons of the filiform sensilla of the seventh segment of the right cercus, indicated by arrows. Each circle shows the location of one filiform sensillum on the ventral surface of segment 7 or 8; the diameter of the circle depends on the length of the hair. (*b*) Drawing of right giant 1 in the last abdominal ganglion. (*c*) Polar plot of the directional sensitivities of the three ventral giants (GI 1–3) that have their axons in the right connectives, constructed as explained in the text. (*a* redrawn after Dagan & Camhi, 1979; *b* modified after Harrow *et al.*, 1980; *c* redrawn after Kolton & Camhi, 1995.)

right cercus from a nozzle that could be rotated to stimulate it from different directions. Air currents were applied from an angle of 45° above the cockroach, which is the type of angle from which an attack might be made. Spikes were counted for the first 50 ms following the start of each stimulus; this is roughly the time over which a giant neuron would respond before a cockroach starts to move in response to an air current. The results are expressed as a polar plot (Fig. 3.13*c*). The origin of the plot represents the cercus, and each response is plotted as a point whose direction from the origin represents the direction of the stimulus, and whose distance from the origin represents the number of spikes.

In Fig. 3.13*c*, each point is the mean of several stimuli in experiments on five different cockroaches. For each interneuron, the different points are joined by a line, and the resulting shape gives an immediate impression of the directional sensitivity of that giant. The shape and location of the plot for each giant are called its **receptive field**, a term that is commonly used to refer to the region around an animal from which a neuron receives sensory input. All three interneurons respond preferentially, but not exclusively, to stimuli coming from the same side as their axons (the right side in this case), and interneuron 1 responds well to stimuli from the front or from the rear. There is a little overlap between the receptive fields of interneurons 2 and 3, but if we look at the receptive field to determine which stimulus direction elicits the best responses in these two neurons, we find that the directions for the two giants are almost exactly at right angles to each other: interneuron 2 prefers stimuli from behind and to the side of the animal, whereas interneuron 3 responds best to stimuli from the front and the side. In fact, the best direction for interneuron 3 corresponds to the direction in which the cercus normally points, and the best direction for interneuron 2 is at right angles to this. Thus, a cockroach can determine whether an air current is coming from the left or from the right by comparing the responses of its left and right giants; and it can distinguish stimuli coming from the front from those coming from the rear by the relative excitation of interneurons 2 and 3. If a current of air is coming directly from the side, these two giants will respond with almost equal vigour.

A cockroach giant interneuron can generate 25 spikes or more between the times when it starts to respond to wind and when the cockroach starts to move. By stimulating single giants to generate extra spikes during responses to wind, Liebenthal, Uhlman & Camhi (1994) showed that the most critical parameter in the response is the number of spikes in left compared with right giants. It is, therefore, not possible to equate a single spike, or any other single event, with a distinct decision to initiate an escape turn in the same way as a spike in a Mauthner neuron or crayfish lateral giant marks the decision to initiate the startle responses. No single giant interneuron responds exclusively to air currents from the left or from the right, yet the cockroach always makes quite accurately oriented turns that take it away from a source of danger. Control of the activity in leg motor neurons must, therefore, involve quite sophisticated mechanisms to combine signals that originate in different giant interneurons (Levi & Camhi, 1996).

3.11 Conclusions

Behaviour is initiated as a result of a decision-making process in the central nervous system. Startle behaviour provides simple and dramatic examples of this, which illustrate how it is possible to trace the orderly flow of information through pathways of interconnected neurons. These examples enable us to understand how a specific behaviour pattern is triggered and executed, and so provide some general lessons about the role of nerve cells in behaviour.

Startle behaviour usually begins with the stimulation of specific groups of sensory neurons, and these in turn excite one or more orders of interneuron, which filter the sensory input. The filtered information is then integrated by interneurons that act as a switch or trigger, determining whether or not that particular behaviour is initiated. In the first two examples discussed here, this executive function is concentrated in a single giant neuron, and this yields an extremely rapid response. But in cockroach escape, the executive function is spread more diffusely over a group of interneurons and this permits a more finely tuned response.

One important aspect of the way these neuronal circuits work is that both inhibition and excitation play a vital role. The pattern of inhibitory connections is essential for ensuring that the behaviour is executed efficiently. This is seen clearly in the way that motor systems that would generate movements that are incompatible with a startle response are inhibited, such as the inhibition of the extensor system during flexion of the crayfish abdomen and the mutual inhibition between left and right Mauthner neurons. A second role for inhibition is in switching off sensory systems that are likely to be stimulated by movements originating from the startle response itself. This prevents undesirable consequences such as triggering fresh startle responses and habituation in sensory pathways. Finally, it is common for an excitatory circuit that triggers a startle response to inhibit itself, after a delay, which prevents fresh activity from being triggered until the initial movement has gone to completion. The effectiveness of this inhibitory control is often ensured by having inhibitory input at two or three separate points in a given neural pathway.

Another important aspect of these neuronal circuits is that decisions are made in the nervous system by the integration of synaptic potentials in passive cell membranes. This is clearly illustrated by the way in which the crayfish lateral giant and the Mauthner neuron collect and sum synaptic potentials that arise from sensory input on their dendrites. At each point in the system, it is the properties of the synaptic potentials that determine the outcome. The output of each neuron in a circuit depends on the way in which inputs to it are combined.

Although they play a crucial role in the decision-making process, the giant interneurons described above do not directly drive all the subsequent events making up the escape behaviour. They excite only a small number of neurons that are involved in co-ordinating motor activity; subsequently, other neurons are activated by excitation from other, more slowly conducting interneurons and by feedback pathways from sense organs. The full behaviour pattern is thus produced by co-ordinated activity in many parts of the central nervous system, only some of which are directly driven by triggering interneurons. This arrangement results in a behaviour pattern that can greatly outlast the original sensory response that triggered it. Startle behaviour is unusual in the simplicity of the decision-making process and in the way that, once triggered, it completely overrides other behaviour. But it is these features that have made it possible to bridge the gap between sensory input and motor output. In much neuroethological research, this gap seems to be unbridgeable, at least using currently available techniques.

Further reading

Eaton, R.C. ed, (1984). *Neural Mechanisms in Startle Behavior.* New York:Plenum Press. This is a collection of several articles on startle behaviour in many different types of animals; it is a good review of the state of research at the time it was written.

Faber, D.S., Korn, H. & Lin, J-W. (1991). Role of medullary networks and postsynaptic membrane properties in regulating Mauthner cell responsiveness to sensory excitation. *Brain Behav Evol* **37**, 286–97. This paper reviews the physiology of the sensory input side of the Mauthner neuron. Other papers on this neuron are in the same issue of the journal.

Hoy, R.R. (1993). Acoustic startle: an adaptive behavioural act in flying insects. In *Biological Neural Networks in Invertebrate Neuroethology and Robotics*, pp. 139–58 ed. R.D. Beer, R.E. Ritzmann & T. McKenna, New York: Academic Press. A favourite subject for neuroethologists for many years has been the way many insects can react to the calls of hunting bats.

4 Capturing sensory information

4.1 Introduction

Most animals are active organisms and need up-to-date information about their environment if they are to behave appropriately. Much information is potentially available in the many forms of energy and chemicals that impinge on the surface of the organism and act as stimuli. An animal must be able to detect the various forms of energy and to sort them all out, a job that is carried out by its sense organs, which act as instruments monitoring stimuli coming in from the environment. Sense organs are thus an animal's mechanism for gathering up-to-date information, and as such it is hard to exaggerate their importance in behaviour.

Clearly, a monitoring instrument is useful only if it measures one particular form of stimulus; having an instrument that responded indiscriminately to all forms of stimuli would be nearly as uninformative as having no monitoring facility at all. Hence, one of the most fundamental properties of sense organs is **selectivity**. Each sense organ contains specific **receptor cells** that are tuned to be sensitive to one particular stimulus. In many sense organs, the receptor cells are also nerve cells, having axons that convey their information to the central nervous system, and receptor cells of this kind are, therefore, often called **sensory neurons**. In other sense organs, such as the eyes of vertebrates and insects, the receptor cells do not have long axons themselves but instead make synaptic contact with separate nerve cells that send trains of spikes to the central nervous system.

The particular form of stimulus to which a sense organ responds determines its sensory modality. The diversity of sensory modalities has been recognised since the time of Aristotle, who distinguished the five primary senses: sight, hearing, touch, smell and taste. These five are still the ones recognised in everyday life, but many more modalities have been discovered by physiologists, including unfamiliar ones such as the ability to detect electric or magnetic fields. However, sensory modalities can be separated broadly into three basic categories. First, there are **chemoreceptors**, sense organs that are stimulated by chemical ions or molecules in various forms, either as gases or in solution. These include not only the more familiar senses of smell and taste but also less conspicuous receptors monitoring such things as oxygen or carbon dioxide concentration. Second, there are **mechanoreceptors** that are stimulated by some form of kinetic energy. These include many sense organs that monitor internal functions, such as muscle tension or joint position, as well as senses such as touch, balance and hearing. Third, there are **photoreceptors**, which respond to electro-magnetic energy in the form of photons.

Sense organs falling within these main categories are found throughout the major animal groups. They exhibit a diversity of form as great as that of the animals of which they form a part, even within a single modality. In spite of this diversity, there are some basic receptor mechanisms that are common to all the sensory modalities. Here, the operation of basic sensory mechanisms is illustrated by a simple mechanoreceptor, the campaniform organ of insects.

4.2 Basic receptor mechanisms: the campaniform organ

Campaniform organs detect strains in the cuticular exoskeleton of insects and so provide information about loading on different parts of the body. They occur in groups at those parts where stresses are most likely to be felt, particularly near joints between segments (Fig. 4.1a, b). Mechanoreceptors that detect the relative positions of, or stresses and strains in, different parts of the body are known collectively as **proprioceptors**. The muscle receptor organ in the crayfish abdomen (see section 3.4) is another example of a pro-prioceptor, and the way that a proprioceptor influences the output from motor neurons often fits well the description of feedback in a control system (see section 1.6). The campaniform organs that occur near the joints of cock-roach legs have been studied in detail, and play a significant role in normal walking movements. They detect the strains that occur when the leg is being applied to the ground and used to thrust the animal forwards. The response of the campaniform organs initiates a reflex that prevents the leg being lifted and swung forwards while it is still bearing a load (Zill & Moran, 1981).

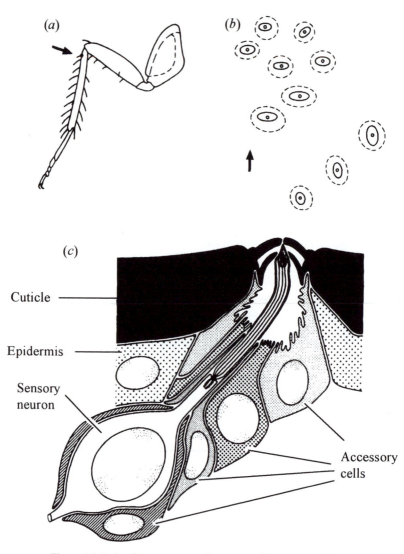

Figure 4.1 A simple sense organ: the campaniform organ of insects. (*a*) A leg of a cockroach (*Periplaneta*), showing the location of the groups of campaniform sensilla (arrow) on the upper surface of the tibia. (*b*) The surface of the tibia viewed from the direction of the arrow in (*a*), showing the distribution of the caps of the campaniform organs. The arrow indicates the longitudinal axis of the leg, pointing towards the body. (*c*) Transverse section through a campaniform organ, showing the relation of the sensory neuron to the cuticular cap and to the accessory cells, as revealed by electron microscopy. (*a* and *b* redrawn after Zill & Moran, 1981; *c* redrawn after Gnatzy & Schmidt, 1971.)

The ability of a campaniform organ to detect strains arises from its specialised construction, which consists of a single sensory neuron coupled with certain accessory cells and structures (Fig. 4.1c). Opposite the axon, the sensory neuron bears a short process, the dendrite, with a specialised ending consisting of a modified ciliary structure. This structure contains numerous microtubules and is connected to a specialised region of cuticle called the cap. Seen from the external surface, each cap is oval in shape, having a central dome and a surrounding depressed region. In section, it can be seen that the cuticle of the cap is thinner than elsewhere and that the centre of the dome fits tightly around the tip of the sensory neuron. It seems clear from this structural arrangement that the cap must act as a mechanism by which the strains in the cuticle are transmitted to the ending of the sensory neuron – it must act as a mechanism for coupling the stimulus energy to the receptor cell.

The electrical activity of a campaniform organ can be studied using the kind of arrangement shown in Fig. 4.2a. A cockroach leg is secured and extracellular electrodes are arranged to record activity in the main leg nerve. When bending forces are applied to the leg with a probe, a campaniform organ responds with a vigorous discharge of spikes (Fig. 4.2b). The same response can be obtained by applying a much finer probe, with a tip diameter of only 1 to 2 μm, directly to the cap of a single campaniform organ. Furthermore, if that particular cap is ablated with a sharp needle, the discharge is no longer recorded in response to a bending force applied to the leg, which confirms the role of the cap in coupling the stimulus to the receptor cell.

The sensory neuron acts as a biological transducer, converting mechanical energy that bends the leg into electrical energy that generates the spikes. This basic process of **transduction** is a defining feature of all receptor cells. The precise mechanism of transduction of the sensory stimulus is only partly understood in most cases, mainly because receptor sites are small and the events occur rapidly. But it is not hard to see that in a mechanoreceptor like the campaniform organ, mechanical distortion of the receptor cell membrane caused by the stimulus could open ion channels in the membrane. This will allow particular ions to pass through the cell membrane, resulting in a change in membrane potential. In campaniform organs, transduction probably occurs in the ciliary region of the dendrite. The dendrite lies in a lumen which almost certainly contains a special

(*a*)

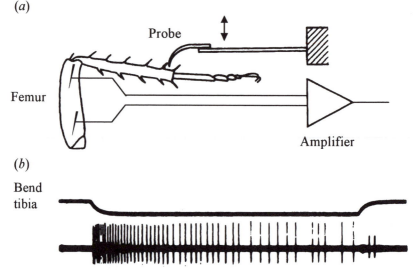

(*b*)

Bend
tibia

Figure 4.2 Physiological recording from the tibial campaniform organs. (*a*) Arrangement for recording spikes in response to controlled stimuli. The stimuli are applied to the distal end of the tibia with a probe attached to a piezoelectric crystal, which bends when a voltage is applied to it. Two fine pins act as electrodes for recording from the leg nerve, and are attached to an amplifier before the signal is displayed on an oscilloscope screen. (*b*) Response to bending the tibia in a ventral direction. The upper trace is a record of the voltage applied to the crystal to bend the tibia, and the lower trace is the extracellular recording of the sensory spikes. (*a* redrawn after Spinola & Chapman, 1975; *b* modified after Zill & Moran, 1981.)

high-potassium solution, a feature that is commonly found in mechano-sensitive organs, including mammalian ears. The accessory cells are probably responsible for regulating the composition of this lumen.

The change in membrane potential that occurs at the site of transduction is known as the **receptor potential**. In another mechanoreceptor of the cockroach leg, a special spine near the knee, the receptor potential has been measured directly with an intracellular microelectrode (Basarsky & French, 1991), but less direct means have been used to record campaniform organ receptor potentials. The result of such an experiment is shown in Fig. 4.3, where the record of membrane potential has been redrawn as if it were recorded intracellularly. In this example, the fine stimulus probe applied to the cap was driven sinusoidally, which is an effective way to stimulate a receptor cell because the amplitude and frequency of the sine wave are

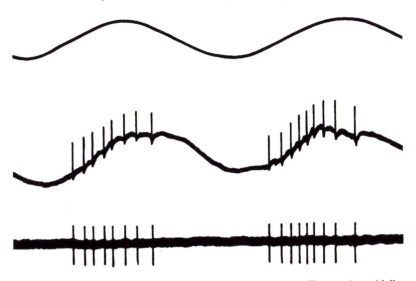

Figure 4.3 The response of a single campaniform sensillum to sinusoidally modulated indentation of its cap with a fine probe. The top trace is the monitor applied to the probe, with upward movement indicating increased force. The bottom trace shows spikes recorded with pin electrodes, as in Fig. 4.2. The middle trace indicates the appearance of the receptor potential as if it had been recorded with an intracellular electrode; upward movement of the trace indicates depolarisation of the cell membrane. This trace has been redrawn from a record of the receptor potential recorded extracellularly. (Modified after Mann & Chapman, 1975.)

readily varied and the resulting responses lend themselves to quantitative analysis.

The sinusoidal stimulus elicits a large receptor potential, which follows the imposed force closely: as the force indenting the cap increases, the cell becomes increasingly depolarised; and as the force decreases the cell becomes repolarised (middle trace in Fig. 4.3). Spikes are generated by the depolarisation and are superimposed on the basic shift in membrane potential; they are propagated along the axon to the central nervous system. The change from receptor potential to spikes is termed **coding** because the analogue signal of the receptor potential is converted into the digital code of spike frequency, which is used for long-distance communication in the nervous system. The conversion of receptor potential into spikes occurs at the place where the axon originates from the cell body, the spike-initiation zone. The mechanism by which it occurs is the same as that by which spikes are triggered by summation of excitatory synaptic potentials in other neurons (see Chapter 2). Transduction of an increasing stimulus will cause a greater current flow and so depolarisation of the membrane at the spike-initiating zone will occur more rapidly. This means that the time interval between successive spikes will be less, and so the spike frequency will be higher.

Inspection of Fig. 4.3 shows that, although the receptor potential mirrors the changing stimulus rather faithfully, the train of spikes reflects the stimulus pattern less satisfactorily. The axon must employ spikes for communication because the graded receptor potential could not travel far, being limited by cable properties of the neuron. However, the change from an analogue to a digital code involves a certain loss of information because the use of a frequency code means that both encoding and decoding the signal require time – a particular stimulus strength is immediately represented by a particular receptor potential, but requires at least two separate spikes to encode it. The closer together the spikes are in time, the stronger the stimulus.

In addition to the intrinsic limitations of coding, there is another process at work in shaping the response of the sensory neuron. Looking back to Fig. 4.2b, when a constant bending force is first applied, there is a vigorous response to the onset of the stimulus but spike frequency then soon declines. Similarly, in Fig. 4.3, spike frequency initially rises with increasing stimulation, but begins to decline even before the stimulating force has reached its maximum value. This pattern of response, whereby a receptor cell responds vigorously to a changing stimulus but soon ceases to respond to a steady stimulus, is known as **adaptation** and is a property of most sensory cells. Generally, changes in stimulus strength are more significant than constant stimuli, and adaptation is a means of enhancing the detection of changing stimuli.

A sensory neuron that is much more sensitive to a rapid change in the stimulus is said to be quickly adapting or **phasic**; one that is sensitive to slowly changing or maintained stimulation is said to be slowly adapting or **tonic**. Naturally, phasic and tonic receptor cells provide an animal with information about different types of situation. A phasic receptor provides information about rapidly changing events, whereas a tonic receptor keeps an animal informed about a steady background situation. Many sensory neurons adapt to a stimulus in a way that represents some combination of phasic and tonic properties. The campaniform organ is an example of this because it is most sensitive to a rapid increase in applied force, but it also exhibits a declining response to a maintained stimulus.

The tibial campaniform organs in the cockroach leg are arranged in two subgroups with mutually perpendicular cap orientations (see Fig. 4.1b). The longitudinal axes of the caps of the distal group lie parallel to the longitudinal axis of the tibia. In physiological recordings, it is found that campaniform organs in the distal group respond only to downward bending of the tibia and those in the proximal group respond only to upward bending. This pattern of response can be understood by thinking of the tibia as a cardboard tube and considering the effects of bending in one direction and then in the opposite. The upper surface of the tibia, where the campaniform organs are located, will undergo transverse compression when the tibia is bent down and longitudinal compression when it is bent up. Hence, the fact that the distal campaniform organs respond to downward bending and the proximal ones to upward bending shows that each sensory neuron is sensitive only to transverse compression of its cap (Zill & Moran, 1981).

The tibial campaniform organs, therefore, act as directionally sensitive strain gauges and the two groups of campaniform organs below the knee provide the animal with information about opposing forces to which the tibia is subjected during walking. The organs are most sensitive to bending forces in the plane of the femoro-tibial joint, whether these are produced by external factors or by contraction of the muscles that move the tibia. Forces that bend the leg at unnatural angles or other forces, such as those twisting it about its longitudinal axis, produce little or no response in any of the campaniform organs. The local forces of compression that best excite a particular campaniform organ constitute its **receptive field**, a concept previously introduced in the description of the wind-sensitive giant interneurons of the cockroach (see Chapter 3).

4.3 Summary of basic mechanisms

The basic processes of sensory action, illustrated by the campaniform organ, can now be reviewed by reference to Fig. 4.4. The incoming energy is coupled to the sensory ending by means of highly specialised accessory structures and is transduced in the sensory neuron terminal into a receptor potential. This is a smoothly graded potential which spreads passively

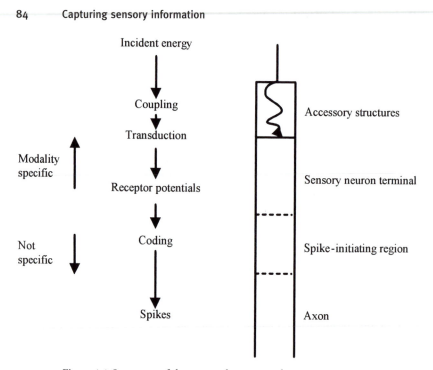

Figure 4.4 Summary of the events that occur when a sensory receptor responds to a stimulus. The normal sequence of processes is shown by the arrows in the centre of the figure, and the structures in which these processes occur are shown on the right. The arrows on the left indicate the extent to which these events are specific to a particular modality (details in text). Where the receptor cell works without trains of spikes, the receptor potential spreads passively down the axon to the synapse with second-order neurons.

along the neuron membrane. When it spreads to a special region at the start of the axon called the spike-initiation zone, the receptor potential is converted into a digital code of spikes. The spikes propagate along the axon towards the central nervous system.

As indicated on the left of Fig. 4.4, the more central processes, such as coding, are not specific to the particular type of receptor. Indeed, the coding of passive potentials into spikes is not specific even to sensory neurons, but is common to all neurons that employ trains of spikes. The more peripheral processes are modality-specific: the transduction process determines the sensory modality, and the coupling process, effected by the accessory structures, narrows down the modality and sharpens the receptive field.

4.4 Essential properties of eyes

The same basic processes that have been illustrated in a simple mechano-receptor underlie the operation of more complicated sense organs. There is no sense organ more appropriate to the study of animal behaviour than one of the arrays of photoreceptors, commonly called eyes. The surface of the Earth is continuously bathed in light, and the speed with which it travels means that light can be a source of almost instantaneous information about events that are occurring at a distance. Hence, it is not surprising that eyes have evolved in many different animal groups, and that vision plays a crucial role in guiding behaviour patterns in many species.

Two fundamental factors influence the design of any sense organ. First, there are the physical constraints imposed by the particular modality involved. In the case of eyes, the physical properties of light set the limits to performance. Features such as the overall size of the eye, the design of lenses, and the arrangement of light-sensing elements will partly be deter-mined by the physical properties of light, just as in any other optical instru-ment. The second factor that influences the design of a sense organ is the need to gather information that is relevant to the animal's particular way of life. Thus, animals that are active at night time will have eyes of a construc-tion different from those of animals that are usually only active in daylight. The design of an animal's sense organs must form an integrated part of its overall adaptive strategy.

Eyes reach their highest level of complexity amongst the vertebrates and arthropods, especially insects, as well as in cephalopod molluscs. The eyes of insects are quite different in their design from those of vertebrates. Nevertheless, both types of eye have to deal with the same problems in con-verting patterns of change in light intensity into signals that are useful for guiding behaviour. The same physical constraints influence the design of the optics of both types of eye, and there are many similarities in the way the two types of eyes operate. In photoreceptors, the same basic sensory processes that have been described in a mechanoreceptor occur. Energy from the environment is focused onto a photoreceptor by specialised accessory structures (lenses) and it is then transduced into a receptor potential, which typically adapts fairly rapidly. Most photoreceptors do not have long axons, and the receptor potential itself regulates the transmission of signals across synapses to the next layer of cells, the second-order neurons. Here, the compound eyes of insects are used to illustrate the principal features of eyes and the properties of the receptor potential.

4.5 Design features of eyes

The lens of an eye gathers light from the environment and usually focuses it as a crisp image onto photoreceptor cells. The greater the amount of light that one photoreceptor receives, the easier it is for it to provide an accurate measure of light intensity. This is particularly important for animals that are active in dim lighting conditions, when the amount of light available is small. There are two main ways of maximising the amount of light that a photoreceptor receives: one is for the eye to have a large lens, which increases the amount of light caught from each part of the visual field; the other is for the photoreceptors themselves to be large, so that each captures light over a relatively large area of the image. But having large photoreceptors is not an advantage if the animal needs to extract detail from the image on its retina, because what is required is for neighbouring photoreceptors to sample small but distinct areas. The greater the number of sampling stations that cover one area of the image, the greater the amount of detail that can be resolved in that area. The way in which an eye is constructed, therefore, involves compromise. The compromise between the need for large photoreceptors, which have good ability to capture available light, and small photoreceptors, to enable the image to be examined in detail, is the most basic. The ability to measure light levels is called **sensitivity**, and the ability to discriminate fine detail in an image is called **resolution**. The two are quite distinct requirements, and it is difficult to improve the performance of an optical instrument for one of them without deterioration in the other.

Insects often depend heavily on vision, which they achieve by means of compound eyes. A compound eye consists of many discrete optical units, each of which has its own lens that focuses light from a small part of the environment onto its own small group of photoreceptors (Fig. 4.5). Each unit is called an **ommatidium**, and its lens is often referred to as a single facet. Large dragonflies have almost 30000 ommatidia in each eye, but most insects have fewer than this – the housefly has 3000 and *Drosophila* has 700. Compound eyes give the animal a very wide field of view; a fly can see in almost every direction from its head. The size of eye that an insect

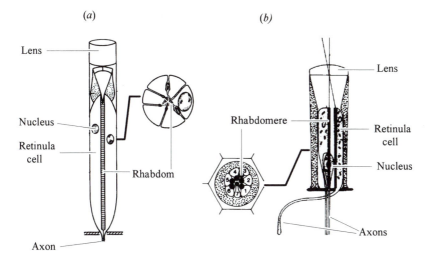

Figure 4.5 The arrangement of cells in the ommatidia of two diurnal insects. (*a*) A locust (*Locusta*) ommatidium. Transverse sections of the ommatidium at two levels (on the right) show how the rhabdomeres of individual retinula cells form a single, central, light-collecting rod, the rhabdom. (*b*) A blowfly (*Calliphora*) ommatidium. Here, the rhabdomeres of different photoreceptors within an ommatidium remain separate. Seven of the photoreceptors are numbered in the cross-section, on the left. Also shown are pigment cells that surround each ommatidium. (*a* from Wilson, Garrard & McGiness, 1978; copyright 1978 Springer-Verlag; *b* from Hardie, 1986; reprinted with permission from *Trends in Neuroscience*; copyright © 1986 Elsevier Science.)

can carry around is obviously limited and, in order to achieve good resolution, the size of each ommatidium must be small so that as many as possible can be packed into the eye to provide a reasonable sampling density. However, as the size of a lens is reduced, the image that it produces becomes increasingly blurred, and there is a limit to how small individual ommatidia can be.

The reason why small lenses produce blurred images is that when light passes through an aperture or is bent by passing from one medium to another, some waves are scattered – a property called **diffraction**. Because of this scattering, light that passes through one part of a lens will spread out and interfere with light that passes through another part of the lens. This mutual interference causes blurring in the image that the lens projects onto its focal plane. The image of a small spot of light is focused by a lens to a diffuse spot called an Airy disk. The width of the Airy disk decreases as the diameter of the lens increases, and so an eye that has a wide lens with a large aperture will be able to work with a sharper image than an eye that has a small lens.

If an eye is to make effective use of a sharply focused image, the width of its photoreceptors must be matched to the width of the Airy disk. If the photoreceptors were wider than the Airy disk, two closely spaced images could not be distinguished from each other. However, interference effects set a lower limit to the width of the **rhabdom**, the photoreceptor element that catches the light which strikes its end, acting as a light guide. As the diameter of a light guide is reduced towards the wavelength of light, interference effects frustrate total internal reflection and an increasing amount of light travels outside the light guide. The parts of photoreceptors that capture light energy and transduce it into an electrical signal are cylindrical light guides. If a large amount of light is lost from these structures, the performance of the eye is degraded because light travelling outside the receptor element will not be captured by the receptor cell and may even cause cross-talk by entering a neighbouring cell. In view of this, light guides made of living tissue cannot usefully be less than 1 μm in diameter. Thus, diffraction sets a lower limit both to blurring of the image and to the size of the receptor elements.

In honey bees, which are typical of insects that fly on bright days, most facets are about 25 μm across, and collect light from a cone with an angle of a little over 1°. To resolve two small spots of light as separate rather than one larger spot, a bee would have to use three ommatidia – one for each spot, with an un-illuminated ommatidium between them. Therefore, bees can distinguish between two objects that are about 3° apart. Compound eyes of species that are active in dim light conditions often have ommatidia that are larger than would be expected to give the best balance between sampling frequency and a sharply focused image (Snyder, Stavenga & Laughlin 1977). Usually, in dim light conditions, photoreceptors also collect light not just from the lens of their own ommatidia, but from surrounding lenses as well, an optical arrangement that is called superposition. Insects and crustacea that are active in bright daylight usually operate with the best resolution that a compound eye can deliver; this type of eye is called an apposition compound eye, and ommatidia are shielded from their neighbours so that each works quite independently.

Some diurnal insects depend on vision for the localisation of prey or sexual partners. A good example is the mantis *Tenodera*, which depends on vision to locate its prey. In order to achieve good resolution for a task such as this, there is a need to dedicate as many ommatidia as possible to sample the visual environment, and each individual ommatidium must have a crisp image of a small part of the environment. This means that each ommatidium must have a lens that is as large as possible. The size that an insect's eye can grow to is clearly limited, however, and there is a compromise between the number of ommatidia and their size in a particular region of the eye. The mantis eye has evolved so that only a small portion of each eye has the high resolution required to examine potential prey. This region, where resolution is enhanced, is called a **fovea**. The mantis fovea is functionally equivalent to the more familiar fovea of vertebrate eyes.

The fovea of a mantis eye is a small region at the front of the eye (Fig. 4.6a). Individual ommatidia here have lenses that are 50 μm in diameter, larger than elsewhere in the eye, and rhabdoms that are 1.5 to 2 μm across, narrower than elsewhere in the eye, a specialisation to maximise resolution by capturing light over a relatively narrow cone of acceptance. Also, the angle between adjacent ommatidia (0.6°) is narrower than elsewhere in the eye (Fig. 4.6b), and so the surface of the eye is less curved here than elsewhere. These structural features are consistent with physiological properties of ommatidia, measured by recording the responses of photoreceptors with intracellular electrodes. Intracellular recordings show that individual ommatidia accept light from a cone of angle 7°, and this is smaller than the acceptance angle for ommatidia measured elsewhere over the eye. In bright light, acceptance angles, measured in electrophysiological experiments, are very close to the angles between adjacent ommatidia, measured anatomically. In dim light, ommatidia in the fovea generate smaller responses than those elsewhere in the eye, which means that they are less sensitive. This is consistent with their narrow rhabdoms. The mantis eye is thus constructed so that resolution is increased at the expense of sensitivity in one region of the eye, the fovea, and is correspondingly reduced in the rest of the eye.

When a prey-like object appears in its peripheral field of view, a mantis responds with a rapid movement of the head that brings the object's image into the fovea (see Fig. 1.7 p. 16). Subsequent movements of the prey are followed by tracking movements of the head, which hold the image of the prey in the fovea while the body is aligned and brought into range for a raptorial

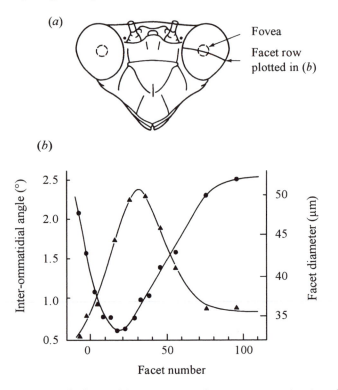

Figure 4.6 The fovea of the mantis (*Tenodera*) eye. (*a*) Anterior view of the head; the broken circle within each eye indicates the region of inter-omma-tidial angles that constitutes the fovea. (*b*) Facet diameters (●) and inter-ommatidial angles (▲) plotted against position in the eye for the row of facets indicated in (*a*). Note how facet diameters increase as inter-omma-tidial angles decrease. The acceptance angles of individual ommatidia follow the curve for inter-ommatidial angles very closely. (Redrawn after Rossel, 1979.)

strike. There is a large area of overlap in the visual fields of the two eyes in the mantis, so that this insect can use binocular cues to estimate prey distance. The foveae of the two eyes are included in the binocular field, and the central axes of the left and right foveae intersect in the sagittal plane about 4 cm in front of the head. It is evident from watching the prey-catching behaviour that the function of the foveae is close examination of spatial detail of potential prey, while the peripheral eye is chiefly responsible for the detection of novel objects.

In absolute resolving power, insect foveae are about an order of magnitude poorer than human peripheral vision and two orders of magnitude poorer than human foveal vision. However, a major function of high resolution is to enable animals to detect objects at a distance. Larger animals need to detect objects at greater distances than small animals. A small insect may need to react to objects that are only a few centimetres away, whereas the appropriate reaction distance for a large primate may be several metres. The force of this point can be made by multiplying the angular separation between visual receptors by the height of the animal, and then comparing the values obtained from a number of active species. Expressed in this way, resolution turns out to be remarkably uniform over a wide range of animals, from *Drosophila* to *Homo*. Thus, compound eyes provide their small owners with resolving power that is as good as that of vertebrates when considered in terms of biological needs rather than physical ideals.

4.6 Photoreceptors and the receptor potential

Within an ommatidium, the lens focuses light onto the tips of photoreceptor cells, which are called **retinula** ('little retina') cells (see Fig. 4.5). There are usually eight or nine retinula cells per ommatidium. A retinula cell is elongated, with its longitudinal axis parallel to that of the ommatidium. Outermost is the cell body, and it includes a specialised structure, the **rhabdomere**, where phototransduction occurs. The cell membrane is folded into very regular finger-like projections, or microvilli, in a rhabdomere. The microvilli contain the photopigment, a rhodopsin molecule, that absorbs light and initiates the process of phototransduction. In most insects, including bees, locusts and dragonflies, the rhabdomeres of all the retinula cells of an ommatidium are adjacent to each other, creating a central rod-like structure, the rhabdom (see Fig. 4.5*a*). In most insects, all the retinula cells of one ommatidium share the same field of view, and all converge on the second-order neurons that are beneath their ommatidium. However, advanced dipteran flies, such as blowflies, have a different structure, where individual rhabdomeres remain separate (see Fig. 4.5*b*), and each photoreceptor of a rhabdom looks in a slightly different direction.

Light has a dual physical nature. The way in which it is diffracted when it passes through lenses and apertures is a characteristic of its wave-like properties. It also consists of discrete packets of energy, called **photons**. The amount of energy in a photon depends on the colour, or wavelength, of its light. Light is actually radiation that occupies a very narrow part of the electromagnetic spectrum, the part that animals can make use of with their eyes. Infrared radiation has a longer wavelength than visible light, and has insufficient energy to trigger phototransduction. At the other end of the visible spectrum, the amount of energy in an ultraviolet ray is so high that it is potentially damaging to biological molecules.

Photoreceptors are capable of responding to single photons of light, if the eye has been left in darkness for some time to maximise its sensitivity. In an insect, each photon gives rise to a discrete, depolarising potential, often called a bump because of its shape. Bumps are usually 1–2 mV high in intracellular recordings (Fig. 4.7a). In absolute darkness, bumps are very rare, and the evidence that each bump represents the absorption of a photon comes from a comparison between the statistical properties of the arrival of photons from a very dim light source at a retinula cell and of the occurrence of bumps in the cell. About 60 per cent of the photons that arrive at a facet generate a bump (Lillywhite, 1977).

In very dim light conditions, photoreceptor potentials consist of a series of discrete photon bumps. As light intensity increases, the bumps fuse together, and the photoreceptor response to a change in light intensity becomes much more smooth (Fig. 4.7b). In daylight, an insect retinula cell receives millions of photons a second. If each bump generated a signal of 1 mV, this would lead to a signal of about 1000 V in the photoreceptor, which it is clearly incapable of generating. When exposed to light, the sensitivity of the photoreceptor response decreases quite dramatically, so that each additional photon generates a smaller response. This is called **light adaptation**. Conversely, adaptation to decreasing levels of ambient illumination, which causes sensitivity to increase, is called **dark adaptation**. Light and dark adaptation enable photoreceptor responses to be appropriate to the average ambient light intensity. A number of different mechanisms are involved, including changes in pigment-containing cells that regulate the arrival of light at a rhabdom (analogous to the pupil of vertebrate eyes), and changes in the sensitivity of transduction. It is important for photoreceptors to be able to adjust their sensitivity to ambient lighting conditions, because these can vary enormously. In sunlight, each photoreceptor receives about 40 million photons per second, and this number decreases to 40 000/s inside a room and 40/s in moonlight. Because natural light

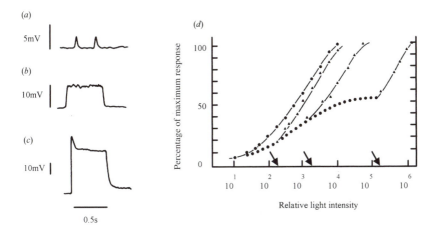

Figure 4.7 Coding of light intensity by an insect photoreceptor cell. (*a–c*) Recordings from a dragonfly (*Hemicordulia*) retinula cell of the responses to three different levels of light stimuli, delivered in darkness. Extremely dim light (*a*) elicits two single photon bumps; moderately dim light (*b*) elicits a steady receptor potential in which small fluctuations, caused by the random nature of photon arrival, can still be discerned. Bright light (*c*) elicits a receptor potential with a sharp, initial peak, followed by a more sustained level. Notice how the voltage calibration changes in (*a*) to (*c*). (*d*) Intensity–response curves for a dragonfly retinula cell. The amplitude of the receptor potential is expressed as a percentage of the maximum response recorded, and light intensity is expressed on a logarithmic scale. The curve on the left (•-•-•) shows the peak size of the receptor potential to flashes of increasing intensity in a dark-adapted cell. The three curves to the right (▲-▲-▲) are plots for cells adapted to particular levels of background illumination, indicated by the arrows. The row of dots connecting the bottoms of the four plots shows the steady-state amplitude of the receptor potential during different levels of sustained background illumination. (Recordings from Laughlin, 1981; copyright © 1981 Springer-Verlag.)

intensity varies so much, it is convenient to use a logarithmic scale to express it. On such a scale, each unit usually represents a tenfold change in light intensity. A change in intensity from 10 to 100 occupies the same space on such a scale as a change from 1000 to 10000 (10 is 10^1, or $\log_{10} 10$ is 1; and 100 is 10^2, or $\log_{10} 100$ is 2, and so on).

One effect of light adaptation is that, at quite moderate light intensity, the response to a step increase in light consists of a sharp, initial depolarising peak, which quickly drops to a more sustained or plateau receptor potential (Fig. 4.7*c*). If the increase in light is maintained, the photoreceptor potential declines very gradually, but stabilises over a timescale of a few minutes. This means that a photoreceptor has both tonic and phasic characteristics. The sustained receptor potential that is produced in response to a long-lasting change in the overall level of illumination is a tonic characteristic, and the initial peak response, which emphasises a change in light intensity, is a phasic characteristic.

The second effect of adaptation is apparent in the way that information about light intensity is encoded as a particular value of receptor potential. This pattern of adaptation is usually examined quantitatively by plotting intensity–response curves, based on intracellular recording of the receptor potential (Fig. 4.7d). At the start of an experiment, the eye is kept in the fully dark adapted state and the sizes of the peak responses to steadily brighter flashes of light are recorded. The time intervals between successive flashes are kept long enough to ensure that the eye remains fully dark adapted throughout the experiment. The non-linear nature of bump summation is reflected in the sigmoid shape of the curve of peak receptor potential against a logarithmic scale of light intensity. For low intensities of stimulus, there is a gradual rise in the size of the receptor potential as stimulus intensity increases. The maximum intensity that the receptor is capable of signalling is about three log units (1000), on the horizontal axis of the graph. At this intensity, the response has **saturated**. However, over a range of intensities of three log units (a thousandfold change in absolute intensity) the size of the response is proportional to the logarithm of light intensity.

The next stage in the experiment is to superimpose the stimulus light on a steady, background level of illumination. The responses to superimposed test flashes are then recorded in the same way as before. Fig. 4.7d shows that background illumination shifts the intensity–response curve along the intensity axis, and the size of the shift depends on the intensity of the background light. At low intensities at which adaptation in the response waveform is small, the shift is also small. However, at higher intensities, the shift produced becomes larger. Hence, the range of intensities to which the cell responds is shifted to match the level of background illumination. This shift in the curves represents a loss in sensitivity: more light is now needed for the photoreceptor to generate the same voltage response. The curves keep approximately the same shape, indicating that the relation between test flash intensity and peak response remains much the same, and response amplitude changes in proportion to the logarithm of stimulus intensity. The extent of adaptation varies between types of eyes, and a very good illustration of this comes from recent research on adaptation in the eyes of different species of dipteran flies. This research shows that the physiology of photoreceptors is well correlated with the ecology and lifestyle of a particular species (Box 4.1).

Box 4.1. Fast and slow photoreceptors

Tipula, a slow-flying, nocturnal cranefly, and *Sarcophaga*, a fast-flying, diurnal fleshfly, are two of the 20 species of dipteran fly surveyed by Laughlin & Weckström (1993). The intracellular recordings in the figure are responses to 0.5s pulses of light delivered to dark-adapted eyes. The receptor potential in *Tipula* rises relatively slowly, and does not adapt during the light pulse, whereas that of *Sarcophaga* rises rapidly and, for the brighter pulses, adapts by decaying quickly from an initial peak to a more sustained level. *Tipula* is more sensitive to light than *Sarcophaga*, as shown by the intensity versus response curves and by the larger bump responses to individual photons in dim light. Photoreceptors of *Sarcophaga* are fast, and able to respond quickly to fluctuations in light as the insect flies around among twigs and leaves. Those of *Tipula* are too slow to help a fast-flying animal avoid colliding with objects, but are good at detecting light and dark at night time. *Sarcophaga*'s photoreceptors are fast because they contain a special potassium channel. The lifestyle of *Tipula* means it does not need fast photoreceptors; in fact, to have them would be a significant metabolic cost to the animal. (Figure modified from Laughlin & Weckström (1993), copyright Springer-Verlag.)

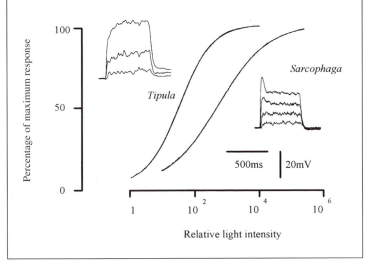

It is a common pattern in sensory receptors for the amplitude of the response to be proportional to the logarithm of stimulus intensity, and this makes excellent functional sense. First of all, the large range of light intensities to which an eye is exposed in the day-to-day life of an animal is compressed into a manageable scale. Second, responding in this way has the effect of making equal *relative* changes in light intensity generate equal *absolute* changes in the size of the receptor potential. One log unit on the light intensity scale is a tenfold change in absolute light intensity, which causes a change in receptor potential of about 35 per cent of its maximum amplitude. A tenfold change in light intensity is a very large change, and a photoreceptor would normally experience much smaller changes in intensity as it scans a natural image. Coding light intensity in this way enables the eye to recognise different objects because they are distinguishable by differences in the proportion of light that they reflect. The relative brightness of two objects is called their **contrast**. Contrasts of objects do not vary when ambient illumination changes. For instance, the contrast between the print and the white paper on this page will be the same whether you are reading the book in a dimly lit room or outside on a beach in bright sunlight. The black print will actually reflect more light on the beach than the white paper will when the book is read in the room. The signal that a photoreceptor generates, therefore, consists of a small, steady depolarising potential, which depends on the mean or background level of light, plus fluctuations of a few millivolts caused by viewing objects of different contrasts.

The receptor potential is not coded into spikes in an insect retinula cell, but is conducted by passive spread along the axon. Sometimes, the axon can be as long as 2 mm, but the cable properties of these axons are such that little of the signal is lost at the synaptic terminals.

4.7 Conclusions

Sense organs are essentially biological transducers, converting incoming energy to the changes in membrane potential that are the currency of information used throughout the nervous system. One of their key features is selectivity. Selectivity starts at the level of the receptor cells, which are essentially biological transducers, converting incoming energy to the membrane potential changes that are used throughout the nervous system. Each sense organ is selective for a particular form of physical or chemical stimulus and so acts as a tuned monitoring instrument. Just as in manufactured instruments, there are physical constraints that set limits to their performance. One of the outcomes of these constraints is that an improvement in one direction cannot be made without detriment in another, which is well illustrated for the compromise between resolution and sensitivity in the design of eyes.

For a sense organ to function effectively as a monitoring instrument, the stimulus energy must be appropriately coupled to the site of transduction in its sensory receptors. Consideration of both campaniform organs and eyes shows that coupling structures are highly specialised to enable a sense organ to operate close to the limit of what is physically possible. The receptive fields of individual sensory neurons are finely adjusted to enable a group of receptors to pass on the most useful information to the central nervous system. Thus, in gathering the sensory information that will regulate an animal's behaviour, many events of crucial importance take place at the coupling stage before any neural signal has been generated.

The result of sensory transduction is the generation of a receptor potential in each of the sensory neurons that have been stimulated. A common feature of sensory receptors is that their sensitivity adapts to particular background levels of stimulation. Adaptation ensures that signals about changes in stimulus strength are enhanced, and also enables receptors to operate efficiently over a wide range of stimulus intensities. In some sense organs, the receptor potential is immediately coded into spikes and the next stage of processing occurs within the central nervous system. In others, notably eyes, the receptor potential itself is conducted passively down a short axon for synaptic communication with the next layer of cells. In either event, it is the array of receptor potentials in the sense organs that constitutes the nervous system's view of the world. All subsequent neuronal processing depends on the receptor potentials; all later integration is built on this initial basis.

Further reading

French, A.S. (1988). Transduction mechanisms of mechanosensilla. *Ann Rev Entomol* **33**, 39–58. An account of different types of mechanosensors in insects.
Nilson, D-E. (1990). From cornea to retinal image in invertebrate eyes. *Trends*

Neurosci **13,** 55–63. A description of the large number of different kinds of compound eyes that are found in crustaceans and insects.

Stavenga, D.G. & Hardie, R.C. eds, (1989). *Facets of Vision.* Berlin: Springer-Verlag. A book containing well-written articles on various aspects of the structure and function of compound eyes; especially relevant to this chapter are the chapters by M. Land and S.B. Laughlin.

5 Stimulus filtering: vision and motion detection

5.1 Introduction

Sensory systems have evolved to provide information that is particularly relevant to an animal's way of life. Sensory neurons have modalities and receptive fields that are strongly biased in favour of gathering information that is behaviourally significant for that species. Whatever their bias, sense organs can pick up large amounts of information about an animal's environment: for example, the photoreceptors of an insect's eye provide a point-by-point representation of light levels in the surrounding visual environment. Higher-order neurons in a sensory system cope with all this information by discarding much of it and keeping only the most significant aspects. These neurons act essentially as **filters**, and transmit only certain aspects of the signal they receive. A consequence of this is that much of the information present at the level of the sensory receptors is thrown away.

Filtering is largely achieved by circuits, in which neurons interact with each other through their synaptic connections. As a result of these interactions, some features of the signal are enhanced and others are discarded at each level in a sensory system. This progressive refinement of the sensory signal begins at the very first synapse, between a sensory receptor and a second-order neuron. Generally, lower-order neurons respond to fairly simple characteristics of stimuli, such as changes in brightness. Higher-order neurons, on the other hand, often respond to particular patterns of stimuli in which information coming from particular groups of sensory receptors is combined together. In fact, much of the complexity of the synaptic interactions in a sensory system is concerned with ensuring that there is a consistent response to a specific stimulus pattern, although the physical attributes of the stimulus may vary. An interneuron that responds to a particular combination of stimuli is known as a **feature detector**, and the specific pattern to which it responds may be termed its key feature. Neuronal circuits that select different features commonly occur in parallel at a given level in a sensory system, with each circuit forming a **pathway** for processing a particular feature.

Feature detection has been studied in most detail in visual systems. The clearest examples are those in which visual processing is involved in initiating a simple behaviour pattern of obvious importance, as in the case of prey recognition in toads (discussed in sections 1.4 and 1.5). Early ethologists invoked the concept of a releasing mechanism to explain how a sensory system filters out a particular key feature to trigger an appropriate behaviour pattern. Study of insects has revealed a rich assortment of neurons that respond to particular types of moving stimuli, and we probably know more about the circuits involving movement-detecting neurons in insects than about those in any other type of animal. For some of these neurons, their role in behaviour is quite clear and we have a reasonable idea of how their selective responses are produced by neuronal processing.

5.2 The insect visual system

The interneurons that process visual information in insects are arranged as a series of neuropiles that lie beneath a compound eye and make up an optic lobe of the brain (Fig. 5.1). Most interneurons connect two layers of neuropile, with their cell bodies and input synapses in one layer and an axon that conducts information towards the brain into the next layer of neuropile. Information, therefore, tends to flow sequentially through different layers of neuropile. But there are also circuits that make lateral connections within one layer, and some neurons convey information outwards, from the brain towards the eye.

The photoreceptors convey a point-by-point representation of the environment to the first neuropile, the lamina. Axons that connect the lamina with the next neuropile, the medulla, cross over one another so that the rear of the medulla is connected with the front of the lamina. The spatial representation of the environment is preserved in an array of columnar elements within the medulla, and the presence of a **topographic map** of the environment is a universal feature of complex visual systems. (Topographic means 'description of place' in Greek.) In each medulla column, there are several tens of types of neuron. It is likely that different types of neuron in a

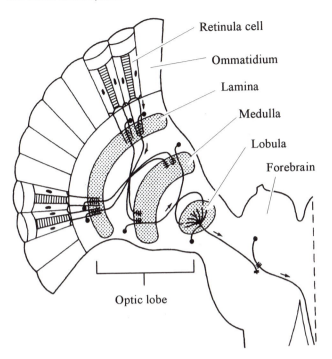

Figure 5.1 The visual system of a diurnal insect, such as a dragonfly or a locust, showing the main neuronal pathways involved in movement detection. The three optic neuropiles (lamina, medulla and lobula) are indicated by light stippling. The small arrows show the pathway followed by information originating in the uppermost ommatidium that is drawn. The general arrangement is similar in flies, except that the photoreceptors of one ommatidium are not fused, and the lobula contains a distinct neuropile called the lobula plate, which would lie on top of the lobula in this drawing.

medulla column are dedicated to filtering out particular stimulus features that occur within the visual field of one ommatidium, or of a group of neighbouring ommatidia. Many of the neurons in the next layer, the lobula, are also arranged in topographic columns. It is hard to make electrophysiological recordings from these columnar neurons in the lobula and medulla because they are tiny. However, some particularly large neurons also occur in the lobula, and stable recordings can be made from them so that their response properties can be studied in detail. These neurons have fields of view that incorporate many ommatidia, and they abstract information about particular types of movement which can occur anywhere within a very large field of view. These movement-sensitive cells, therefore, provide excellent examples of feature-detecting neurons.

The first neuropile, the lamina, is divided into an array of discrete neural units called cartridges. Each cartridge receives its input from photoreceptors that share the same field of view. In most species, including locusts and dragonflies, this means that a cartridge receives input from photoreceptors of the ommatidium that lies above it. In advanced flies, however, each cartridge receives input from just one photoreceptor in the ommatidium directly above, and from one photoreceptor from each of five surrounding ommatidia (see Fig. 4.5b, p. 87). Each cartridge contains the same complement of neurons, and the set of connections made within a cartridge is so precisely determined and repeated across the eye that the structure of the lamina has been described as crystalline.

The principal cells of each lamina cartridge are called large monopolar cells. As the name suggests, the axon of a large monopolar cell gives rise to just one process. A tuft of short, brush-like dendrites arises from it and, in blowflies, these dendrites make almost exactly 200 discrete anatomical synapses with each connecting photoreceptor (Nicol & Meinertzhagen, 1982). The large monopolar cell process then becomes a smooth axon which, together with the axons of smaller neurons, conveys information from the lamina to the medulla. From each ommatidium, two central photoreceptors, that are tuned to detect ultraviolet and blue light, pass straight through the lamina and terminate in the medulla. Within the lamina, amacrine cells make lateral connections between adjacent cartridges, and there may also be feedback neurons that carry signals back from the lamina to photoreceptors.

5.3 Neuronal coding in the insect lamina

The large monopolar cells are specialised to respond to changes, or contrasts, in the visual signal. Like the photoreceptor cells, they convey signals as small, graded changes in membrane potential and do not generate trains of spikes. They are, therefore, particularly sensitive to small fluctuations in light intensity about an average value. Their responses also depend on the contrast between light received by their own photoreceptors and those of neighbouring cartridges.

Characteristic intracellular responses from a photoreceptor and a large monopolar cell to a pulse of light are shown in Fig. 5.2. Photoreceptors

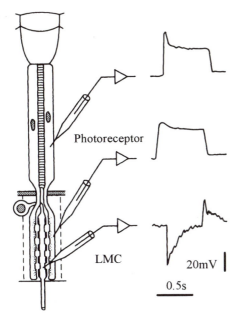

Figure 5.2 The transfer of signals across the first synapse in an insect visual pathway. The drawing on the left shows how the retinula cells of an ommatidium send short axons down to the layer of second-order neurons in the lamina. Intracellular recordings of responses to a 0.5s flash of light were recorded separately from the receptor cell body and axon, and from a large monopolar cell (LMC). (Recordings from Laughlin, 1981; copyright © 1981 Springer-Verlag.)

make inhibitory synapses with large monopolar cells which, therefore, respond to an increase in light with a hyperpolarising signal, a response of the opposite polarity to that of photoreceptors. The response by a large monopolar cell is not, however, a mirror image of the response in a photoreceptor. These second-order neurons respond much more phasically than photoreceptors to a change in light, which is well illustrated in the waveforms of the responses to a pulse of light shown in Fig. 5.2. The response of the photoreceptor shows an initial peak depolarising potential, followed by a sustained, smaller depolarisation that lasts until the end of the stimulus. The large monopolar cell marks the start of the stimulus by a large, transient hyperpolarising signal, and the end of the stimulus by a transient depolarising signal. During the light stimulus, its membrane potential repolarises almost to its level in darkness. Thus, information about mean

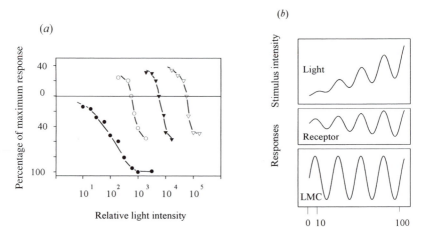

Figure 5.3 Coding of light intensity by large monopolar cells (LMCs) of dragonfly (*Hemicordulia*) or blowfly (*Calliphora*). (*a*) Peak response amplitudes plotted as a function of the intensity of illumination delivered from a small point source of light, centred on the receptive field of a LMC. For the curve on the left, the LMC is dark adapted (●), whereas for the others it is light adapted to various different background intensities. For each curve, background intensity can be read from the intersect with the axis that shows zero response. The responses to both light-on and light-off were measured relative to the largest hyperpolarising response to light-on. (*b*) A schematic diagram to show coding of light-intensity in retinula cells and LMCs. The light signal varies sinusoidally and gradually increases in mean intensity, representing the envelope of light-intensity variations that might be experienced by scanning a visual scene at different intensities of ambient light. Because of adaptation, when the mean light level increases, the retinula cell response does not grow as dramatically as the light-intensity signal. The LMC response is proportional to the contrast in the light signal: most of the signal about ambient light intensity is removed, and the remaining signal is amplified. (Modified after Laughlin & Hardie, 1978.)

levels of illumination is lost in passing from photoreceptors to large monopolar cells, but information about changes in illumination is enhanced. The loss of information about the mean level of illumination is a good example of adaptation in a sensory system.

The way in which the large monopolar cells code visual stimuli has been studied by Simon Laughlin and colleagues in the same way as has been done for photoreceptors (Fig. 5.3). First, in the dark-adapted state, responses are measured to brief light stimuli of different intensities delivered from a dark background. Then a series of intensity–response curves is plotted, each one recording responses to increases and decreases in light from a particular mean, or background, level. Because adaptation quickly returns the membrane potential of a large monopolar cell to near its dark resting potential following a change in light, there is no steady-state response to background illumination. Consequently, the large monopolar cell response to an increase or decrease in light always starts from the same value of membrane potential, and the intensity–response curve moves horizontally by an amount equal to the change in background light intensity. Small increases and decreases in intensity produce, respectively, hyperpolarising and depolarising departures from resting potential.

The size of a response to a particular change in light intensity is always greater in a large monopolar cell than in a photoreceptor. In other words, as the signal is passed across the first synapse in the pathway it is amplified as well as being inverted. In blowflies, the signal is amplified about six times. The relative change in light during a particular stimulus is expressed as its contrast, and so large monopolar cells generate larger responses than photoreceptors to light signals of particular contrasts.

Quantitatively, the relationship between contrast and response can be measured as the gradient of the intensity–response curve; the steeper the gradient, the greater the sensitivity to small changes in light intensity. Large monopolar cells have steeper intensity–response curves than photoreceptors, reflecting their greater sensitivity to small fluctuations in light from a particular mean intensity. As a result, for a particular background level of light, the response by a large monopolar cell to increases in light intensity saturates at a lower intensity than the response by a photoreceptor. Saturation in the response is shown by a levelling-off in the intensity–response curve; it is the intensity at which a further increase in light will no longer lead to an increase in response amplitude.

The range of decreases in intensity that a light-adapted, large monopolar cell can cover is also reduced compared with a photoreceptor. Consequently, there is a trade-off in response characteristics: the price of an increase in sensitivity to changes in light is a reduction in the range of changes in intensity that can be covered. The response by a large monopolar cell to a signal with a particular contrast is the same, irrespective of the background, adapting light level.

When an insect experiences a large change in background light intensity, for example by flying out of dense shade into bright sunlight, its large monopolar cells can be driven beyond the range of their intensity–response curves. However, light adaptation is sufficiently rapid that large monopolar cell membrane potential usually settles to near its original value within less than a second, and the neuron's ability to respond to small changes in light intensity is restored. Following a change in light intensity of 100 times, the initial, large response by a large monopolar cell has almost completely decayed within 200 ms. Very large changes in light intensity such as this are extremely rare in the day-to-day life of an insect, and the intensity–response curve of a large monopolar cell spans the range of changes in intensity that the insect will encounter most often as it moves around.

Thus, the synapses that connect a photoreceptor with a large monopolar cell process the visual signal in two important ways. They filter it, so that information about background intensity is subtracted; and they amplify it, so that the signal about small changes in intensity is transmitted. Amplification, to make the signal as large as possible at an early stage in the visual pathway, is important because every time the signal passes across a synapse from one neuron to another it can become contaminated with noise (Laughlin, Howard & Blakeslee, 1987).

Another kind of transformation that occurs involves mutual inhibition between nearby cartridges. This kind of inhibition between neighbouring units in a sensory system is widespread, and is known as **lateral inhibition**. It is a mechanism for sharpening the receptive field of a neuron so that it responds well to small stimuli centred on its own receptive field, but not to large stimuli that fall on the receptive fields of many neurons. The signal about light levels detected by surrounding ommatidia is subtracted from the signal in the cartridge of a central ommatidium.

Lateral inhibition was first investigated in a compound eye of the horseshoe crab (*Limulus*) and was subsequently demonstrated to occur in vertebrate retinas (Box 5.1). Its action can be shown in an insect large monopolar cell by stimulating photoreceptors with a spot of light that is small enough to illuminate only a single photoreceptor. The spot can be directed at different angles to the ommatidium, and the responses by a large monopolar cell to the light change with the angle of the stimulus (Fig. 5.4). When the light is centred on the central axis of the large monopolar cell's receptive field, the cell's biggest responses are recorded, and the response amplitude

Box 5.1. Processing and filtering in the vertebrate retina
Although the insect eye is quite different in structure from the vertebrate retina, both have the function of converting a physical image
into a neuronal representation that the brain can act upon, and there
are striking parallels in the way the two operate, particularly in the
early stages. The figure summarises some of the principal types of cells
and their signals in response to a short pulse of light directed to the
photoreceptor on the left, based mainly on work on the amphibian
Necturus (see Dowling, 1987). Rod and cone photoreceptors transduce
light into receptor potentials, and signals pass first to bipolar cells and
then to retinal ganglion cells, which have axons that travel into the
brain (about 1.2 million in humans; Sterling, 1998). Signals are
modified by two layers of horizontally oriented cells, horizontal cells
and amacrine cells. Adaptation to changes in background light levels
occurs at several stages, including in the machinery responsible for
transduction in the photoreceptors, in the strengths of electrical
synapses that link photoreceptors, and in the chemical output

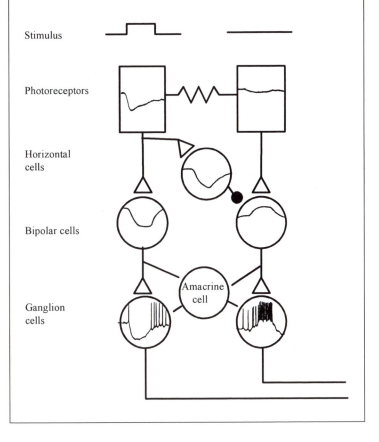

Box 5.1. (*cont.*)
synapses from photoreceptors. Bipolar cells have steeper intensity–response curves than photoreceptors, and these curves remain centred on the mean background light level as it alters so, like the large monopolar cells in the fly eye, bipolar cells signal contrasts in the visual stimulus (Laughlin, 1994). Bipolar cells also have a centre-surround organisation to their receptive fields, in which the action of the photoreceptors directly above a bipolar cell is opposed by the surrounding photoreceptors, an action mediated via the horizontal cells. (Figure redrawn after Dowling, 1970)

decreases as the light is moved off axis. When the spot is shone from about 1.5° off axis, it begins to illuminate neighbouring ommatidia, and the large monopolar cell starts to generate a depolarising rather than a hyperpolarising response. The receptive field of the large monopolar cell, therefore, has two different regions: a central region where stimuli elicit the largest responses (+ sign in Fig 5.4), and a surrounding region that elicits responses with the opposite polarity (− sign in Fig. 5.4). These two regions of the receptive field work in opposition to each other, and the organisation of the receptive field is described as **centre-surround**. This type of organisation is extremely common for neurons in sensory systems. It acts as a type of sensory filter because it enables neurons to react strongly to small stimuli that are centred on their own receptive field, but less strongly or not at all to stimuli that are large enough to enter the receptive fields of several neurons.

A centre-surround receptive field organisation is an efficient way of processing visual information because it enables the size of neuronal signals to be related to the likelihood of a particular stimulus occurring. Two different points in space are more likely to be equally illuminated if they are close together than if they are far apart. This means that if we know how bright one point is, we could predict how bright the second is with a certainty that is related to their separation. In terms of recognising objects, we do not need to know the exact brightness of the two points, but whether they are different from each other – information about their exact brightness is redundant. Careful study of the responses by large monopolar cells in the fly has shown that the strengths of feedback and lateral inhibition are tuned to reduce redundant signals and enhance the detection of contrasts between neighbouring cartridges (Srinivasan, Laughlin & Dubs, 1982).

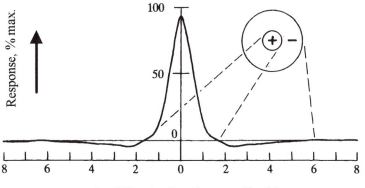

Angle of illumination (degrees off axis)

Figure 5.4 The sensitivity of a large monopolar cell to light incident upon the retina from different directions, expressed as a percentage of the maximum response. Zero response level is the membrane potential in the dark. Through lateral inhibition, off-axis light elicits a small response of opposite polarity to on-axis light, so that the receptive field has an excitatory centre and an inhibitory surround, as shown in the inset. (Redrawn after Srinivasan *et al.*, 1982.)

5.4 Optomotor neurons in flies

The detection of different kinds of movements plays a very significant part in controlling the behaviour of most fast-moving animals, such as flies. These insects are able to make high-speed turns during flight, and are able to land precisely on small targets, like a blade of grass or the edge of a cup. They mate on the wing, and males pursue females under visual control (Land & Collett, 1974). Many of these behaviours are almost certainly under the control of a group of large, fan-shaped neurons that occupy a distinct region of neuropile towards the posterior of the lobula called the lobula plate. There are about 60 of these neurons (Hausen & Egelhaaf, 1989), and many of them are involved in **optomotor** responses. Unlike the large monopolar cells, they are not concerned with stimuli that affect single ommatidia, but they respond to particular movements that occur anywhere within a large receptive field.

Optomotor responses result in movements that tend to keep images of the environment in constant positions on the retina. These simple behaviours play vital roles in enabling more complex behaviour patterns to be executed smoothly. They cause a stationary animal to maintain a constant position and orientation, or help a moving animal to maintain a particular course of locomotion.

An easy way to observe an optomotor response is to place an animal in the centre of an upright cylinder that has vertical dark and light stripes painted on its inner wall. If the cylinder is rotated, the animal will swivel around its vertical axis, following the movement of the stripes. Measurements of optomotor responses in a variety of types of animal have shown that the animal attempts to minimise slippage of the images of the stripes over the retina. Because most of the visual field of the animal is occupied by the stripes, the animal interprets movement of the drum as movement of itself relative to the environment, and optomotor responses naturally tend to stabilise the animal's eyes.

Flies reliably exhibit optomotor responses during flight. The strengths and directions of these responses can be measured in the laboratory by gluing a holder to the fly's back and suspending it in the air. When the fly's feet lose contact with the ground, it beats its wings up and down as if flying, and it will attempt to twist around its holder when it sees vertically oriented stripes drifting past its eyes. The strength of the twisting force that the fly applies to its holder can be measured, and the direction of the force changes whenever the direction of travel of the stripes is reversed (Fig. 5.5*a*, *b*). This kind of movement occurs in the yaw plane, and will tend to correct the animal's flight path when it twists around its vertical axis. Similar optomotor responses also operate in the roll and pitch planes (Fig. 5.5*c*). A twisting movement by the fly, such as yawing, is characterised by a difference in the direction in which images move over the two eyes, whereas a translating movement, such as flying straight forwards, is characterised by images that move over the two eyes in the same direction and at the same speed. When an animal moves relative to its environment, a particular pattern of direction of movements is detected by the eyes and that pattern depends on the type of movement. During forward movement, for example, images move from the front of the animal towards the rear. The slowest movements will be detected by the parts of the eyes directed towards the front of the animal, and the fastest movements will be detected by the parts of the eyes directed towards the sides of the animal. On the other hand, when a flying animal pitches downwards, a rotating pattern of movements is set up, with the front part of each eye receiving upward motion and the back part receiving downward motion. In Fig. 5.5*d* speed and direction of movement

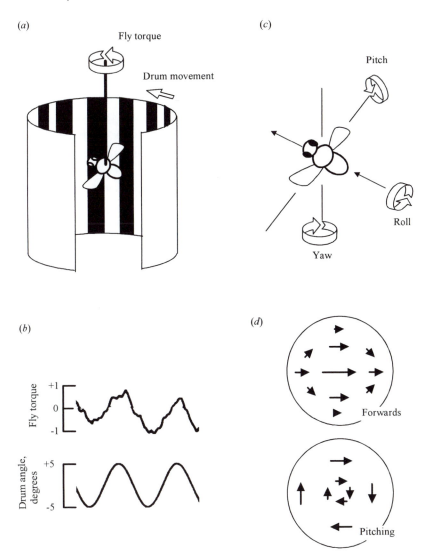

Figure 5.5 Optomotor responses and flow fields. (*a*) A fly, such as *Calliphora* or *Musca*, is suspended in the centre of a striped drum that can be rotated. The wire tethering the fly is attached to a meter that measures the torque, or turning force, produced by the fly. (*b*) Recordings of drum movements and torque produced by a fly. (*c*) The three axes for rotatory and translatory movements that an airborne fly can make. (*d*) Flow fields that stimulate the right eye when the insect flies forwards or pitches downwards. (*b* recordings from Egelhaaf, 1985; copyright © 1985 Springer-Verlag.).

of images over local regions of the eye are indicated by the lengths and directions of the arrows. The pattern of coherent motion associated with a particular kind of movement is called a **flow field**, and an animal can gather information about its movement through the environment by detecting the kind of flow field that its eyes are receiving.

Many of the large, fan-shaped neurons of the fly lobula plate are unique individuals, and generate their largest responses to particular types of stimulus movement. They filter out particular stimulus features and ignore others. The exact location of a moving stimulus is not significant to these neurons, but direction of movement is, and all these neurons show particular **direction selectivity**. For example, the three HS neurons respond most strongly to images that rotate around the animal in the horizontal plane, and the 11 VS neurons respond to movements upwards or downwards (Fig. 5.6a). Each HS neuron in the right lobula plate is most strongly excited by movements occurring in the clockwise direction around the animal (backwards over its own eye or forwards over the other eye), and those in the left lobula plate are most excited by anticlockwise movements. The most effective direction for exciting a neuron is called the **preferred** direction. Often, neurons are inhibited by movement in the opposite direction, which is called the **null** direction. Each VS neuron is tuned to respond most strongly to a particular type of flow field (Krapp & Hengstenberg, 1996). One VS neuron, for example, responds most strongly to one direction of rolling about the longitudinal axis, but other VS neurons are tuned to detect twisting movements that pitch the head upwards or downwards. The VS neurons, therefore, provide an array of filters, in which a particular combination of pitch and roll during flight corresponds to the strongest excitation of one of the VS neurons. Abstraction of particular stimulus features is often achieved by an array of neurons like this, each of which is tuned slightly differently from its neighbours.

Although the HS neurons have no branches that extend across the brain into the opposite optic lobe, they respond to movements over both eyes. They are most sensitive to stimuli that rotate about the animal in one direction, being excited by stimuli that move backwards over their own eye or forwards over the opposite eye. Neurons such as H1 (Fig. 5.6b) are responsible for carrying information from one lobula plate to the other, and provide pathways that enable the fly to compare stimuli that affect the two eyes. These pathways enable the fly to distinguish whether it is rotating or

(a)

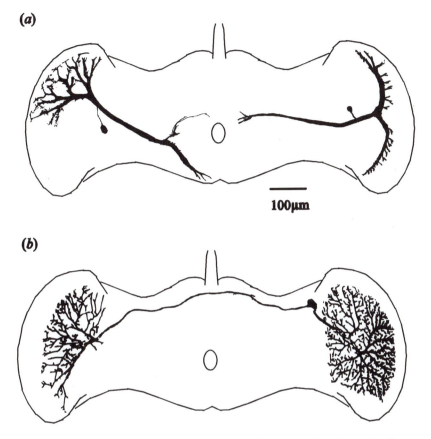

100μm

(b)

Figure 5.6 Optomotor neurons in the lobula plate of a blowfly. (a) On the left is one of the three HS neurons that would respond to anticlockwise movements around the yaw axis; and on the right is one of the VS neurons that responds to downward motion over its eye. (b) Neuron H1, which is excited by movements in the back-to-front direction over its own eye, the right eye for the H1 drawn here. Its axon carries information from one lobula plate to the other. (a modified from Dvorak, Bishop & Eckert, 1975; copyright © 1975 Springer-Verlag; b modified from Franceschini et al., 1989; copyright © 1989 Springer-Verlag.)

moving forwards. During rotation, movement flows forwards over one eye and backwards over the other, whereas when the fly moves forwards, both eyes see backwards movement.

The ways that the HS and VS neurons respond to moving stimuli suggest that they could be important in controlling optomotor responses and there is strong, indirect evidence that they play such a role (Hausen & Egelhaaf,

1989). The output processes of the HS and VS neurons intermingle with dendrites of a small group of neurons that carry information from the brain to motor neurons that control flight and neck muscles. Optomotor responses are impaired in flies after minute cuts are made into the brain to sever the axons of these neurons, or if the HS neurons are killed by irradiating them with a laser. Also, in *Drosophila*, some mutants that lack the HS neurons do not produce normal optomotor responses. Therefore, it is likely that these neurons are an important part of the cockpit of the fly, helping it to maintain a particular flight course.

5.5 Figure-ground neurons of the lobula plate

While an insect is flying around, it needs to detect and react to nearby objects as well as to monitor the way the background moves. A fly, for example, tends to chase other flies, and to select suitable targets such as twigs on which to land. When a fly approaches a twig, the image of the twig will move more rapidly over the fly's eyes than will the images of more distant objects. Tethered, suspended flies turn strongly towards a small object that is moving backwards relative to them, and this response is enhanced if the object is moving over a background pattern that is also moving, rather than remaining stationary (Kimmerle, Warzecha & Egelhaaf, 1997). Likely candidate neurons for recognising a small object and initiating a turn by the fly towards it are a group of four figure-ground neurons in each lobula plate (Egelhaaf, 1985). These neurons select a figure, or object, that is distinct from the general background. Like the HS neurons, they are excited by movements backwards over the eye. They also have large receptive fields but, unlike the HS neurons, they respond much more briskly to movements of small targets than to large parts of the visual field. This is because they receive a balance of excitatory and inhibitory synaptic inputs in response to moving stimuli (Warzecha, Egelhaaf & Borst, 1993). A figure-ground neuron responds strongly to a stimulus that extends over about 5° of the visual field, but progressively less strongly to stimuli that are seen by larger regions of the eye.

5.6 A mechanism for directional selectivity

In sensory systems, there are many examples of neurons that act as filters, responding most vigorously to one particular stimulus configuration. The selectivity of these neurons for their preferred stimulus must arise from the way in which their presynaptic neurons are arranged, and study of fly lobula plate neurons has enabled insights into the way in which one particular kind of selectivity – directional selectivity – arises. The neurons are considered to be driven by an array of local motion-detecting units, each of which detects the movement of images in a particular direction over a small part of the eye. These units are commonly referred to as **elementary motion detectors**. Their properties are known in some detail, although the neurons of which they are composed have not been identified. Many of the properties of elementary motion detectors were first discovered in the 1950s from experiments on the optomotor behaviour of beetles and flies, well before methods had been developed for recording signals from single neurons in the fly brain. More recently, their properties have been inferred by recording the responses by lobula plate neurons to moving stimuli that are seen by small regions of the eye, when the number of local motion detectors stimulated would be very small.

When the image of a moving object travels across an eye, different photoreceptors are stimulated in a particular order. Information about the direction of movement could, therefore, be obtained from the sequence in which the photoreceptors are stimulated. The open rhabdom design of the fly eye has enabled Nicholas Franceschini and colleagues (Franceschini, Riehle & Le Nestour, 1989) to perform a remarkable experiment in which directionally selective responses from a lobula plate neuron were elicited by stimulating just two photoreceptors in sequence (Fig. 5.7a). The neuron they chose to study was H1 (see Fig. 5.6b), which is the easiest lobula plate neuron to record from. A special optical instrument enabled an experimenter to view a single ommatidium and to direct a tiny spot of light onto each of two individual photoreceptors which view adjacent points in space. H1 is excited when the posterior receptor is stimulated just before the anterior receptor, and it is inhibited when the two receptors are stimulated in the reverse order (Fig. 5.7b). This corresponds with the directional selectivity by H1 for moving objects, which is forwards over its own eye.

Illumination of either receptor alone, or both at the same time, does not cause any response in H1; there must be a delay between stimulation of one and stimulation of the other. Illumination of the posterior receptor **facilitates** a response by H1 to stimulation of the anterior receptor. This facilitation does not start immediately when the posterior receptor is stimulated, but after a

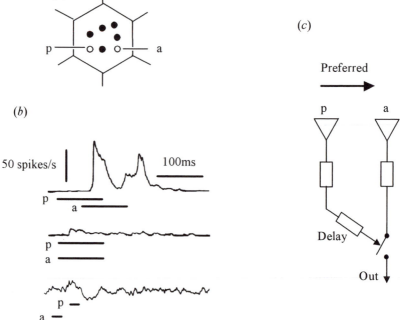

Figure 5.7 Directional selectivity in the housefly (*Musca*) H1 neuron to stimulation of two individual photoreceptors. (*a*) A fly ommatidium viewed through its lens, showing the seven individual rhabdomeres that are visible. The anterior (a) and posterior (p) photoreceptors that were stimulated in the experiment are indicated. (*b*) Responses by H1 to different sequences of stimulating the two photoreceptors. The times when a pinpoint of light was directed at each photoreceptor are indicated. There was a vigorous response when a stimulus to the posterior receptor preceded that to the anterior, but almost no response when the two receptors were stimulated simultaneously. Note that the light-on and light-off stimuli generated separate peaks in the response. In another experiment (shown in the bottom trace), stimulation of the anterior receptor before the posterior one (the null direction) caused inhibition of H1. Each recording was an average from 100 stimulus repetitions, and shows the response of H1 as instantaneous spike frequency. (*c*) Diagram of an elementary motion detector. Open triangles represent the two photoreceptors. (*b* modified from Franceschini *et al.*, 1989; copyright © 1989 Springer-Verlag.)

delay of at least 10 ms. It reaches maximum strength after 50 ms, and lasts for up to 250 ms. This means that H1 is most sensitive to movements that stimulate the anterior receptor 50 ms after the posterior receptor (these movements are produced by images travelling over the surface of the eye at a speed of 72 degrees per second). The delay in facilitation ensures that H1 is not excited when overall illumination of the eyes alters or if light intensity flickers, which would occur when the insect flies into or out of shade.

The wiring diagram in Fig. 5.7c summarises how a directionally selective motion detector, with a preferred direction from posterior (p) to anterior (a), might be organised. The diagram summarises how information is processed. The inputs are elements that respond to light-on or light-off stimuli, and each probably includes a photoreceptor and various neurons in the lamina and medulla. The signal from the posterior photoreceptor is delayed before it is combined with that from the anterior photoreceptor. The combination is indicated as a switch that enables one photoreceptor to regulate the output of the other, although in reality the way the outputs from the two photoreceptors is combined is more sophisticated than a simple on–off device. The two signals are thought to be multiplied together, which ensures that the output is very small when only one of the two receptors is stimulated (multiplying by zero gives a result of zero). The result is that a strong output from the motion detector is only produced when stimulation of the second photoreceptor follows stimulation of the first photoreceptor with a certain delay.

It would be reasonable to assume the medulla contains many such circuits, perhaps one per column, so that H1 is excited by movement in its preferred direction over any part of its eye. In fact, at each location there must be four copies of similar circuits. The first duplication is necessary to account for inhibition of H1 by stimuli moving backwards over the eye, the null direction. Inhibition is apparent in experiments in which H1 shows quite a high rate of spontaneous spike discharge in the absence of any movement stimulation, and sequential stimulation of the two photoreceptors in the null direction briefly inhibits H1, reducing its spike rate. This inhibition shows the same time-dependent properties of facilitation as excitation by movement in the preferred direction. To account for this, there must be two mirror-image, motion-detecting circuits working together, one exciting H1 and the other inhibiting it. The second duplication, in which there are two of each excitatory and inhibitory circuit, is

(a)

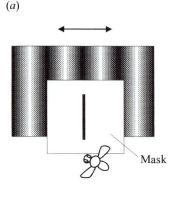

(b)

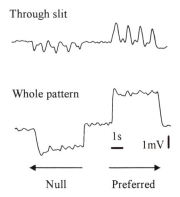

Figure 5.8 Responses by elementary motion detectors. (*a*) When a fly viewed a moving pattern through a vertical slit cut in a mask, only a narrow band of ommatidia and elementary motion detectors was stimulated. (*b*) A wide-field HS neuron responded to each lightening or darkening of this band, as shown in the intracellular recording. When the stimulus moved in the null direction, each response by the strip of motion detectors caused hyperpolarising potentials; and when the stimulus moved in the preferred direction, another set of motion detectors responded, causing depolarising potentials. When the mask was removed so the eye saw the whole moving pattern, responses from different strips of motion detectors summed so that sustained excitatory or inhibitory responses were recorded from the HS neuron. (*b* modified after Egelhaaf & Borst, 1993.)

necessary because light-on stimuli are processed separately from light-off stimuli. Although photoreceptors are excited by light-on stimuli and inhibited by light-off stimuli, neurons such as H1 respond to moving light or dark objects, so signals from photoreceptors must be channelled through different, parallel pathways before they reach the lobula plate. Microstimulation of single pairs of receptors shows that light onto the first receptor facilitates responses to light on, but not to light off, delivered to the second. This duplication ensures that a local motion detector will respond only when there is correspondence in the stimulus that the two photoreceptors see.

In another type of experiment, Martin Egelhaaf and Alexander Borst have made intracellular responses from an HS neuron (Egelhaaf & Borst, 1993). They stimulated the eye with a pattern of vertically oriented stripes that drifted backwards or forwards over it (Fig. 5.8*a*). The darkness of each stripe varied sinusoidally, so there were no sharp borders between light and dark in the pattern. A mask with a narrow slit cut into it was placed between the moving pattern and the eye so that only a narrow, vertically oriented band of ommatidia was stimulated. When the stimulus moved, the photoreceptors in this band experienced sinusoidal changes in light intensity, rather than the abrupt switches in light that occurred during the microstimulation experiments. When the stimulus moved in the preferred direction, the response by the HS neuron consisted of a series of discrete depolarising potentials in which a large potential alternated with a smaller one (Fig. 5.8*b*). Each large potential corresponded with a darkening of the photoreceptors in the band of stimulated ommatidia, and each small potential corresponded with a lightening of this band. When the stimulus moved in the null direction, the neuron responded with a series of alternating large and small hyperpolarising potentials. When the mask was removed so that a large part of the eye saw the moving pattern, the intracellular recording showed a sustained depolarisation of the neuron, indicating excitation, or a sustained hyperpolarising potential, indicating inhibition, depending on which direction the stripes moved. These results are consistent with the hypothesis that motion is detected by local circuits, similar to Fig. 5.7*c*. When the eye sees a large, extended pattern rather than a narrow slit, an HS neuron will receive signals from many vertical bands of ommatidia. Because the outputs from the different bands are out of phase with each other, the summed signal in the HS neuron will be a steady depolarising or hyperpolarising potential in which the fluctuations in output from individual motion detectors are ironed out.

Neurons like H1 and the HS neurons are excited most strongly by movement of large-field stimuli in a particular direction. The amount of excitation also depends on the speed and repeat pattern of the stimulus. These two stimulus features, speed and repeat pattern, cannot be distinguished from each other by an elementary motion detector because the response to a narrow series of stripes moving quite slowly over the eye is the same as the response to stripes that are twice as broad but moving twice as rapidly (Fig. 5.9). The strength of excitation, therefore, depends on the frequency with which the detector is stimulated with changes in the contrast of light. Slowly moving, closely spaced stripes will stimulate a detector with the same contrast frequency as more rapidly moving, broader stripes. The fact that the responses of H1 and HS neurons also depend on contrast

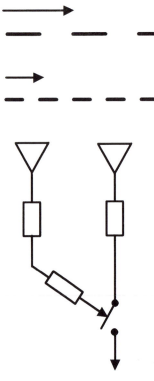

Figure 5.9 An elementary motion-detector circuit cannot distinguish between narrow stripes moving slowly and wider stripes moving more quickly. In this circuit, each receptor is excited whenever a light–dark edge moves over it. The amount by which the detector is excited depends on the delay between excitation of the two receptors, which is short if either the speed of movement is fast or if the stripes are spaced close together.

frequency indicates strongly that they are driven by this type of elementary motion detector. The strengths of optomotor turning responses by a tethered fly also depend on contrast frequency of stimuli, rather than speed or spatial pattern, and this lends support to the hypothesis that the HS neurons are responsible for controlling these behaviours. Some movements by insects, however, are controlled in a way that suggests the insects can measure stimulus speed independently of image structure (Box 5.2). Visual systems channel information about movements through a number of independent pathways, each focusing on particular aspects of the stimulus. Ambiguities, such as that between stimulus velocity and pattern, can be removed at later stages by combining the outputs of different channels.

Box 5.2. Bees can measure image speed to fly a straight course

Bees can be trained to fly along a tunnel if it is part of the route between their hive and a good food source. They tend to maintain a course straight along the centre of the tunnel, and they could do this by balancing the relative speed of motion that is detected by the left and right eyes. Kirchner & Srinivasan (1989; see also Srinivasan,1992) showed that bees detect the speed of motion over each eye by observing flight paths along tunnels with side walls decorated with vertical stripes. No matter how broad the stripes on each side are, the bee flew straight down the middle (*a*). If the stripes on one wall moved (short arrow) in the same direction as the bee, the bee flew closer to that wall (*b*), but if the stripes moved in the opposite direction, the bee flew closer to the other wall (*c*). These results suggest quite strongly that bees have neurons that can compare the speeds of the images of the two walls, something that would be difficult to achieve with the optokinetic neurons of the fly lobula plate.

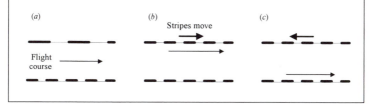

5.7 Summary of fly optomotor neurons

Motion over the eye is first detected by small circuits, called elementary motion detectors, that respond to the sequence in which a pair of photoreceptors is stimulated. Each retinotopic column of neurons in the medulla probably contains several of these detectors, each tuned to a different direction of movement or of light-intensity change. In an elementary motion detector, the signal that originates in one of the photoreceptors is delayed, and is then combined with the signal from the second photoreceptor. The detector, therefore, correlates the signal in one receptor with the signal that occurs slightly later in the second. The detectors drive an array of fan-shaped neurons in the lobula plate, each of which is tuned to a particular direction of motion occurring anywhere in a very large visual field. Most lobula plate neurons respond best to movement of a large panoramic image, and are well suited to controlling optomotor responses that stabilise the image of the environment on the animal's eyes. A few respond selectively to movements of small objects, and these probably elicit turning responses by the fly towards small objects such as other flies or perches.

5.8 Collision warning neurons in the locust

Optomotor responses help to stabilise an animal when it is stationary or proceeding along a straight course, and they do this by referring to movements that occur in the background. The animal also needs to be able to respond to individual objects, such as potential predators, mates or perching sites. Like many animals, locusts and grasshoppers will escape from rapidly approaching objects. This is apparent to anyone who has tried to catch one of these insects; they respond to approach by a powerful jump, caused by rapid extension of the hind legs. Often, the wings are opened during the jump, and the animal extends the range of its jump by flying or gliding.

A particular neuron in the hindbrain of the locust has been found to respond vigorously to the images of approaching objects. The axon of the neuron travels to the thorax, where it excites motor neurons and interneurons that are concerned with the control of the jumping and flying. Probably, therefore, it plays a vital role in channelling sensory information about an approaching object to the motor control circuits responsible for executing escape movements.

It is relatively easy to record responses from this neuron by using extracellular electrodes placed around a nerve cord because it produces spikes that are usually much larger than those from the axons of other neurons. Because the axon of the neuron crosses the brain, and responds to movements detected by the opposite (contralateral) eye, the neuron is called the descending contralateral movement detector (**DCMD**). The DCMD receives its input from a single, large, fan-shaped neuron in the lobula, called the lobula giant movement detector (LGMD; Fig. 5.10). Spikes in the DCMD follow those in the LGMD one for one. The two neurons are extremely sensitive to small movements anywhere within the visual field of their eye, and respond with a brisk, but brief, burst of spikes. As is to be expected from neurons that receive their input through large monopolar cells in the lamina, the response to a sustained stimulus rapidly disappears and they respond to changes in contrast over a wide range of background light intensities.

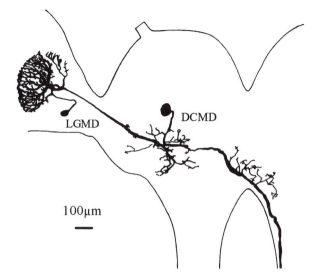

Figure 5.10 Two feature-detecting neurons in the visual system of a locust (*Locusta*), the lobula giant motion detector (LGMD) and the descending contralateral motion detector (DCMD). The drawings were made from neurons that had been stained by injecting them with cobalt ions. The LGMD receives excitatory input from lower-order interneurons on the fan-like array of dendrites on the left. (Modified from Rind, 1984.)

The LGMD generates a vigorous and prolonged train of spikes in response to an approaching object. The frequency of spikes increases throughout the approach movement, as if the neuron has locked on to the approach movement (Fig. 5.11*a*). Images of receding objects, or of objects that are moving around the locust, generate only brief responses. Deviations as small as 2–3° from a direct collision course result in a reduction in response in the LGMD by a half, so it is remarkably tightly tuned to respond to objects that will collide with the animal (Judge & Rind, 1997). The intensity of response depends on the speed with which the object is approaching, and the neuron would respond very well to a predatory bird swooping towards a locust. Wide field movements, like those that cause optomotor responses, inhibit responses by the LGMD.

To determine the types of movement that are most effective at exciting the LGMD, Claire Rind made recordings while a locust viewed a video of a space movie (Rind & Simmons, 1992). This approach was an effective way of providing a wide variety of visual stimuli and rapidly indicated that the

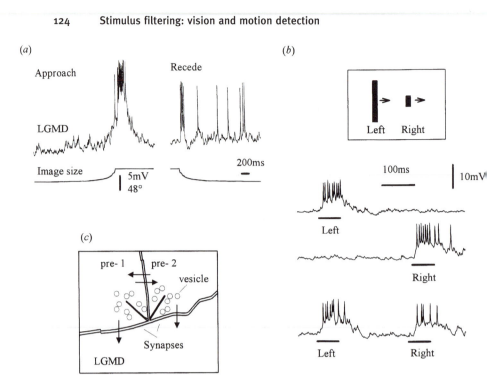

Figure 5.11 Detection of approaching objects by the locust LGMD neuron. (*a*) Response by the LGMD to the image of an object approaching and then receding at 3.5 m/s. The top trace is an intracellular recording from the LGMD, and the lower trace is a monitor of the size of the object's image, which was displayed on a screen. (*b*) Stimulation of one part of the visual field inhibits the response to stimulation of another part. The drawing shows the stimulus screen, with two vertically oriented bars that could move from left to right across the screen. Movement of either bar alone across the screen caused EPSPs and spikes in the LGMD, but when the left bar moved just before the right one, the response to movement of the right bar was greatly reduced. (*c*) Drawing of an electron micrograph of a dendrite of the LGMD with input synapses onto it from two adjacent presynaptic elements, pre-1 and pre-2. Each presynaptic element contains a densely staining bar-like structure that is believed to direct vesicles of neurotransmitter to their site of release at an insect synapse. The LGMD receives synapses from each presynaptic element, and the two elements also synapse with each other as indicated by the arrows. (*a* recordings from Rind, 1996; reprinted with permission of the American Physiological Society; *b* from Rind & Simmons, 1998.)

LGMD responded more vigorously to images of approaching objects than to other types of movement. Subsequently, responses were analysed in more detail by generating carefully controlled moving images with a computer, and then determining exactly which features in the image of an approaching object are the important cues (Simmons & Rind, 1992). Although many invertebrates show avoidance reactions when a shadow is cast over them, the LGMD only responds weakly to overall decreases in light intensity. To excite it, an image of the edges of an approaching object must move over the surface of the eye. Two different features of the image are important as cues that an object is moving nearer to the eye: first, the edges grow in length; and, second, the edges accelerate as they move across the retina.

As with the lobula plate neurons in the fly, selectivity by the LGMD for a particular kind of moving stimulus is established by circuits among neurons in the medulla. The way that these circuits work has been established by using a variety of experimental approaches. Electrophysiological recordings demonstrated that lateral inhibition operates among the columns of neurons in the medulla (O'Shea & Rowell, 1976). When one area of the retina is stimulated by a moving image, the response to stimulation of other parts of the retina is depressed (Fig. 5.11b). Although the first movement clearly decreases the response by the LGMD to the second movement, no inhibitory postsynaptic potentials are recorded from the LGMD. This means that the inhibition must occur at an earlier stage, presynaptically to the LGMD.

Using electron microscopy, it has been shown that the dendrites of the LGMD are covered with input synapses, probably from neurons that originate in the medulla. These synapses are arranged in a remarkable manner (Fig. 5.11c). Each neuron that synapses with the LGMD also synapses with its neighbours; in other words, the neurons that drive the LGMD are reciprocally coupled to each other. These microanatomical circuits provide a route for local lateral inhibition among the elements that excite the LGMD, which is a key feature of the input circuitry to the LGMD, allowing it to filter out approaching objects. In the lamina, the lateral inhibition between cartridges serves a different function – to sharpen the detection of edges in the image.

The way in which the LGMD filters out the image of an approaching object can be envisaged as a kind of race between excitation and inhibition

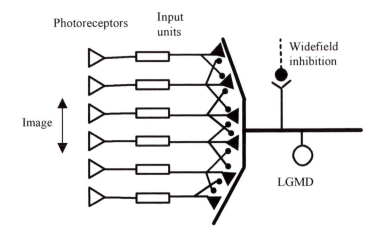

Figure 5.12 A diagram of the circuits thought to drive the LGMD. An array of units drives the LGMD through excitatory inputs (—◄) and inhibits their neighbours. Each unit is excited by a small number of photoreceptors (▷—) and inhibits its neighbours (—●). When an object approaches the eye, its edges spread outwards (arrows), and this creates a race between excitation of units by edge movement, and inhibition between neighbours. The LGMD is excited strongly when excitation is winning the race. A second input to the LGMD causes inhibition in response to wide-field movements.

of the units in the medulla that synapse with the LGMD (Fig. 5.12). The excitation comes from photoreceptors that are stimulated sequentially by the edges of the image as it travels over the eye; and the inhibition travels laterally between the medulla units. As the number of excited units increases, the strength of the lateral inhibition also increases. Consequently, in order for the excitation to outstrip the inhibition and be passed on to the LGMD, a large number of new medulla units must next be stimulated. For this to occur, the extent of the edges in the image or the speed with which they move must be increasing. A network like this has been modelled in a computer, and it responds as predicted, being excited most strongly by the images of approaching objects (Rind & Bramwell, 1996). The network incorporates a second kind of inhibition that also occurs in the LGMD system, and this acts directly on the LGMD itself, causing IPSPs in it. This inhibition acts to reduce responses by the LGMD when the whole background moves, or when overall light intensity changes.

5.9 Conclusions

In a sensory system, the receptors transmit signals to a series of interneurons and, together, these constitute a neuronal pathway in which signals are progressively modified as they are transmitted from stage to stage. Through their synaptic interactions, the interneurons act as selective filters, enhancing some aspects of the received signal and discarding others. A complex sense organ with many receptors is associated with a series of neuropiles made up of many parallel pathways, each processing information from a small group of receptors. Separate classes of interneuron within each pathway extract different types of information from the same receptor input. The neuronal pathways are arranged in an orderly manner, with each one maintaining a particular position relative to its neighbours, so that the central nervous system contains a map of the receptor array and hence, in a visual system, of the external world.

Early in the pathway, interneurons act as signal conditioners, providing a clear and unambiguous signal for the filtering networks that lie downstream. This is illustrated by the large monopolar cells of the insect lamina, in which the signal provided by photoreceptors is processed so that it is amplified and information about background light intensity is discarded. The signals received by interneurons in the medulla are, as a result, largely about contrasts in the light signal – changes both in time and in space. The kinds of operation that medulla interneurons perform are illustrated by the elementary motion detector in flies, or the array of laterally connected neurons thought to drive the LGMD in locusts. We have some good indications of the types of computation that these circuits perform, although direct experimentation, using microelectrodes to record from and stimulate different elements in these circuits, is extremely challenging.

Large neurons in the insect lobula are good examples of feature detectors. They abstract information about a particular aspect of the stimulus, usually the direction of movement, but discard others, such as the location of a target. Neurons of the lobula plate in flies, and the LGMD in the locust, are tuned quite tightly to respond most briskly to particular types of movement. Information about other aspects of a stimulus is processed by other neurons in the visual system, so that a particular scene in the environment is analysed by parcelling different aspects into an array of different interneurons, each concentrating on a different aspect. Understanding the manner in which a nervous system resynthesises a visual scene, combining together information from different feature-detecting pathways, provides a major challenge to neuroscientists.

Further reading

Franceschini, N., Pichon, J.M. & Blanes, C. (1992). From insect vision to robot vision. *Phil Trans R Soc Lond B* **337,** 283–94. This describes how principles concerning vision, derived from the study of motion-detecting neurons in flies, have inspired the design of artificial eyes used to guide the movements of robots.

Laughlin, S.B. (1989). The role of sensory adaptation in the retina. *J Exp Biol* **146,** 39–62. A useful review of sensory adaptation, especially in the insect retina.

Strausfeld, N.J. (1976). *Atlas of an Insect Brain.* New York: Springer-Verlag. A classical description of the different types of neuron found in the brains of flies, particularly in their visual systems.

6 Hearing and hunting: sensory maps

6.1 Introduction

The interaction between a predator and its prey represents a dramatic example of animal behaviour in which the capabilities of nervous systems are stretched to the limit. A hunting animal faces the fundamental problems of detecting and localising the prey, and it must solve them on the basis of purely passive information given out inadvertently by the prey. This is a formidable task and it has led to the evolution of some remarkably sophisticated neuronal systems in species that are adapted for hunting.

If one is asked to name a hunting species, the natural choice is a suitably complex animal such as a large cat or a hawk. These animals do, indeed, possess central nervous systems with the necessary sophistication to handle the complex task of tracking prey, but this sophistication makes most birds and mammals unsuitable as subjects for neuroethological research. However, the difficulty can be overcome by looking at species with a highly specialised method of hunting, based on a sensory system that is dedicated to the specialised method of prey detection and localisation. It then becomes easier to correlate the properties of particular neurons in that system with the particular behavioural task (see section 1.2).

Such dedicated systems are found in two groups of animals that employ hearing as a means of tracking prey, namely owls and bats, which use specialised auditory systems to hunt at night when visually guided predators are at a disadvantage. Owls are able to locate small animals on the ground by listening for the tiny rustling noises made by an animal moving among fallen leaves and twigs. On most nights, an owl's hearing is used in conjunction with its excellent eyesight, but on very dark nights some species can hunt by sound alone.

Insectivorous bats both hunt and find their way around exclusively by sound, using a method that is akin to human sonar and is technically known as **echolocation**. This method involves the bat emitting loud pulses of sound and then analysing the returning echoes in order to find out what lies ahead. Echolocation is a characteristic of bats in the suborder Microchiroptera, a diverse group that is quite distinct from the non-echolocating fruit bats, the Megachiroptera (cf. Macdonald, 1984).

The basic properties of sound and their variation over time determine what is possible by way of detection and localisation for a hunting owl or bat. Sound is created when air molecules are set in motion by a vibrating structure such as a loudspeaker. The vibrations of the speaker generate alternating waves of compression and rarefaction of the air, which propagate out from the speaker at the speed of sound. The molecules involved in propagating the sound move back and forth from regions of high pressure into regions of low pressure, which thereby become regions of high pressure, and so on.

The pressure generated by a sound wave is technically expressed as **sound pressure level** and is measured on a logarithmic scale. The unit most often employed in studying animal sounds is the decibel (dB), which is equivalent to an increase or decrease of about 12.2 per cent in relative sound pressure. Absolute pressure levels are described relative to a reference sound pressure, usually 20 μPa ($= 2 \times 10^{-5}$ N m^{-2}), which is roughly the threshold of human hearing to 1 kHz sound. The interval from a given point on one sound wave to the equivalent point on the next sound wave is the wavelength, which is usually expressed as a **frequency** (the reciprocal of the wavelength) and measured in cycles per second (Hertz). Most sounds that animals produce or listen to have frequencies of thousands of cycles per second, or kiloHertz (kHz).

Neuroethological research on owls has concentrated on the owl's ability to localise a sound source in space by measuring these basic properties of sound. Most research on bat echolocation has paid more attention to the bat's ability to determine the distance from which an echo has returned and to the suitability of different bat sounds for different sonar techniques. Taken together, these studies on bats and owls are providing valuable insight into the neuronal basis of some sophisticated behaviour patterns.

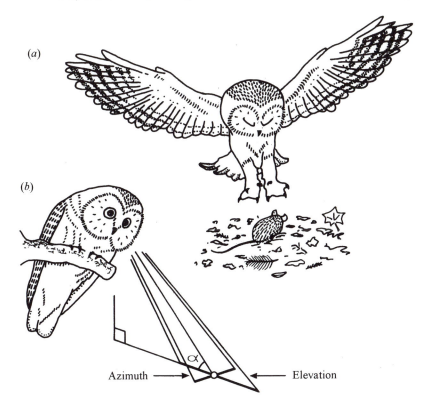

Figure 6.1 The hunting technique of an owl, drawn from photographs of Tengmalm's owl (*Aegolius funereus*) in its natural habitat. (*a*) The owl about to strike prey with its talons, after flying down from an observation perch. (*b*) The owl on its perch immediately before striking, with a diagram showing the errors involved in localising prey by hearing. The prey (O) is observed at a shallow angle (α), with the result that a given angle of error converts into a greater distance along the ground for a vertical (elevation) error than for a horizontal (azimuth) error. (Modified after Norberg, 1970, 1977.)

6.2 Prey localisation by hearing in owls

Adult owls hunt within a well-defined territory, which they know well and patrol regularly at night. During its patrol, an owl visits a number of observation perches, from which it can survey the ground round about. If it hears potential prey, the owl swiftly turns its head so that it directly faces the object of interest. Then, after adequate scrutiny, it flies down to capture the prey in its outspread talons (Fig. 6.1). An owl listening from a perch or in low-level flight must be able to pinpoint the sound source both in the horizontal plane (azimuth) and in the vertical plane (elevation). In fact, unless an owl is looking down from directly above its prey, its orientation is more critical in the vertical than in the horizontal plane (Fig. 6.1*b*).

The use of hearing in prey capture has been studied mostly in the ubiquitous barn owl, *Tyto alba*. Early behavioural studies were carried out using tame individuals, which proved able to locate prey even in total darkness, and a number of simple experiments demonstrated that hearing is used to accomplish this (Payne, 1971). For example, it was found that the birds could strike accurately at a concealed loudspeaker quietly broadcasting a recording of leaf-rustling noises. This result not only showed the barn owl's ability to locate prey by hearing alone, but also opened the way for testing which features of the sound are important in localisation.

Using this technique, the accuracy of localisation was found to vary with sound frequency. The greatest accuracy was achieved with a sound containing frequencies from 6 to 9 kHz. Typically, birds are not sensitive to frequencies above 5 kHz, but the barn owl can hear up to 10 kHz, and 6–9 kHz is the range to which its ear is most sensitive. In addition, more than half the auditory neurons in a barn owl's ear are devoted to this extended frequency range of 5–10 kHz. The usual pattern is for approximately equal numbers of auditory neurons to be devoted to each doubling of frequency (octave), but the barn owl devotes a disproportionate number to the higher frequencies (Koppl, Gleich & Manley, 1993). Hence, these important frequencies can be analysed in greater detail and this arrangement may be thought of as an acoustic fovea (cf. section 6.9 on bats).

The accuracy of sound localisation has been studied in more detail by exploiting the natural response in which an owl turns to face a novel sound. The owl is trained to remain on its perch and the angle through which its head turns in response to a sound is measured using an electromagnetic angle detector (Fig. 6.2 *a*). In each test, the head is first aligned by attracting the owl's attention with a sound from the zeroing speaker, and then the owl is stimulated with a sound from the target speaker.

When the target speaker is placed in front of its face, the barn owl's localisation is exceptionally accurate, with an error of less than 2° in both azimuth and elevation. But the owl's accuracy deteriorates in both planes as the angle between the source and the axis of the head increases (Fig. 6.2 *b*). The rapid flick of the head, with which the owl responds, is initiated about

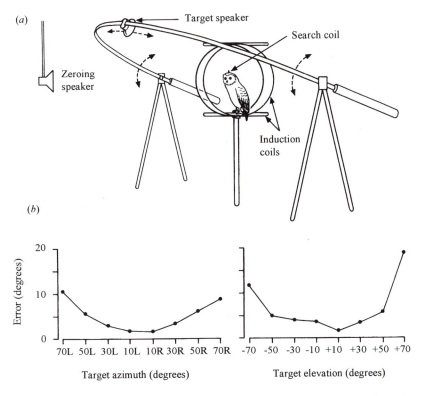

Figure 6.2 Orientation of the head to sounds by the barn owl (*Tyto alba*). (*a*)
The method used to measure the accuracy with which the owl locates sounds
coming from different positions in space. Sound stimuli originate from either
a fixed source (the zeroing speaker) or a movable source (the target speaker).
The search coil on top of the owl's head lies at the intersection of horizontal
and vertical magnetic fields generated by the induction coils. Movement of
the head in response to sound from a speaker induces a current in the search
coil, which is analysed by computer to give horizontal and vertical angles of
movement. (*b*) Localisation accuracy as a function of the position of the
target speaker, showing the mean degree of error in judging target position in
the horizontal plane (left) and in the vertical plane (right) for an individual
owl. (Modified after Knudsen, Blasdel & Konishi, 1979.)

100 ms after the onset of the sound. However, maximum accuracy can be
achieved even with brief sounds (75 ms duration) that end before movement of the head begins. This shows that the owl does not locate the sound
by successive approximation but can determine the sound's precise location in space without feedback. That is to say, the owl is operating under
open-loop conditions (see section 1.6).

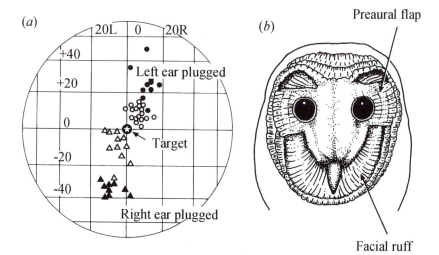

Figure 6.3 The barn owl's ability to locate sounds in elevation. (*a*) A plot of auditory space in front of the owl in degrees of azimuth (L and R) and of elevation (+ and −). The symbols show individual errors in open-loop localisation of a sound source in the centre of the plot (Target) produced by partly blocking one ear. A tighter ear plug (closed circles and triangles) produces a greater localisation error than a looser ear plug (open circles and triangles). (*b*) The facial structures of the barn owl that contribute to localisation of sound in elevation: the facial ruff is formed from tightly packed feathers projecting from the relatively narrow skull, and the ear openings are located behind the preaural flaps. These structures are revealed by removing the sound-transparent feathers of the facial disc, which give the owl's face a flat appearance. (Modified after Knudsen & Konishi, 1979.)

An indication of just what cues are involved in locating a sound is provided by partially blocking one ear, which effectively reduces the sound pressure level at that ear without significantly altering the sound's time of arrival. A plug in one ear leads to significant errors in elevation but only slight errors in azimuth (Fig. 6.3*a*). A plug in the left ear causes the owl to direct its head above and a little to the right of the target, and a plug in the right ear results in the owl facing below and a little to the left. The tighter the ear plug, the greater is the degree of error. This result indicates that the intensity difference between the ears is the principal cue for locating a sound in elevation.

The owl is able to use a comparison of intensity between the left and right ears to locate a sound accurately in elevation due to the arrangement of feathers on its face. The ear openings and the protective, preaural flaps are vertically displaced, the left flap being above the midpoint of the eye and the right one below it (Fig. 6.3b). There is also a slight asymmetry in the facial ruff, which is composed of dense, tightly packed feathers and forms a vertical trough behind each ear opening: the left trough is orientated downwards and the right one upwards. Because of its dense feathers, the facial ruff acts as an effective sound collector at frequencies above 4 kHz. Consequently, the asymmetries in the ruff and in the ear openings give rise to a vertical asymmetry in the directionality of the two ears: the left ear is more sensitive to high-frequency sounds from below and the right ear from above the horizontal plane. If the ruff feathers are removed, the owl is quite unable to locate sounds in elevation and always faces horizontally regardless of the true elevation of the source, but it can still locate in azimuth.

When a source of sound is not directly in front of or behind the head, sound will reach the two ears at slightly different times. In order to test the importance of these time differences, sound stimuli are delivered using miniature earphones installed in the owl's ear canals, rather than sound delivered from a distant speaker, which would inevitably produce differences in both time and intensity. With the earphones, stimuli can be made equal in intensity but different in time of arrival at the two ears. An owl responds to such a stimulus by turning its head horizontally in a direction that the time difference would represent if it were an external source of sound. The owl turns its head to the side that receives the stimulus earlier, and the angle of turning is positively correlated with the magnitude of the time difference between the ears (Moiseff & Konishi, 1981). These experiments show that the owl depends mainly on differences in time of arrival at the two ears to locate sounds in azimuth.

6.3 Auditory interneurons and sound localisation

The fact that a barn owl can localise sounds accurately under open-loop conditions implies that its auditory system can recognise each point in space from a unique combination of the cues that specify azimuth and elevation. Interneurons that respond to such complex combinations of stimulus features are found among higher-order neurons in the midbrain. The main auditory region of the midbrain in owls is situated on the inner edge of the optic tectum (Fig. 6.4a) and is divided functionally into an inner and

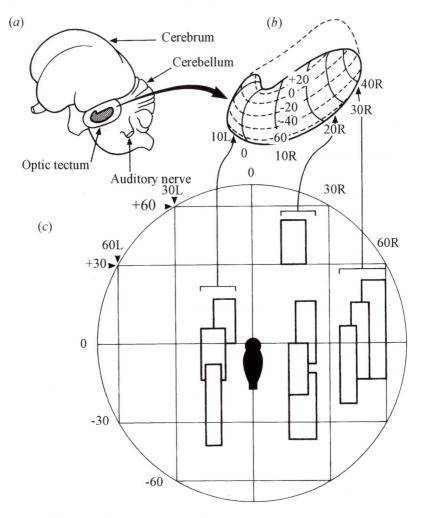

Figure 6.4 The neuronal map of auditory space in the midbrain of the barn owl. (*a*) The left side of the owl's brain, showing the location of the auditory midbrain (in bold outline) on the inner side of the optic tectum. (*b*) The left auditory midbrain enlarged from (*a*), with the co-ordinates of the neuronal map indicated in degrees of azimuth (L and R) and of elevation (+ and −) of auditory space. (*c*) A plot of auditory space in front of the owl in degrees of azimuth and of elevation, showing the receptive fields (bold rectangles) of ten neurons recorded in three separate electrode penetrations. The penetrations were made with the electrode parallel to the transverse plane at the positions indicated by the arrows linking (*c*) to (*b*). (Modified after Knudsen, 1981.)

outer portion. In the inner portion, the interneurons are tuned to particular frequencies and are arranged according to their best frequency, which is termed a **tonotopic** arrangement. The great majority of these neurons respond to their best frequency regardless of where the sound source is located in space.

However, in the outer portion of the auditory midbrain, there are interneurons that respond quite differently; all have similar best frequencies near the upper end of the owl's range (6–8 kHz) and respond only when the sound originates from a specific region of space. These are therefore called space-specific neurons. When recording from such a neuron with a microelectrode, the size and shape of the specific region to which it responds, termed the receptive field, can be determined with the movable speaker (Fig. 6.2*a*). Typically, the neuron is excited by sounds coming from a region of space shaped like a vertically elongated ellipse. Unlike most auditory neurons, the space-specific neurons are insensitive to changes in intensity, and even a 20 dB increase in intensity has little effect on the size of the receptive field.

As the recording electrode is advanced through the outer portion of the auditory midbrain, it samples neighbouring interneurons, which are found to have receptive fields representing neighbouring regions of space. In fact, the space-specific neurons are arranged systematically according to the azimuth and elevation of their receptive fields, so that they form a neuronal map of auditory space (Fig. 6.4*b*, *c*). Two-dimensional space is mapped on to the auditory midbrain, with azimuth being arrayed longitudinally and elevation being arrayed dorsoventrally. On each side of the brain, the map extends from 60° contralateral to 15° ipsilateral, but a disproportionately large number of neurons is devoted to the region between 15° contralateral and 15° ipsilateral. This arrangement means that the 30° of space directly in front of the owl is analysed by an especially large population of neurons on both sides of the brain. In elevation, the map extends from 40° upward to 80° downward but the majority of neurons are devoted to the region below the horizontal (Fig. 6.4*c*).

Partially blocking one ear, which changes the normal intensity differences between the ears, causes a significant vertical displacement of the receptive field in the space-specific neurons but makes little difference in azimuth. Changes in time differences between the ears can be delivered with the miniature earphones, and it is found that individual neurons respond only to a narrow range of time differences, which correspond with the horizontal location of the neuron's receptive field. Neurons that respond to small time differences have receptive fields in front of the face and those that respond to larger time differences have receptive fields at greater angles to the face. This shows that the azimuthal position of the receptive field is determined largely by time differences between the ears (Moiseff & Konishi, 1981).

Thus, the acoustic cues that are used to create the receptive fields of these auditory interneurons are exactly the same cues as the intact owl depends on for sound localisation. This fact makes it almost certain that the neuronal map of auditory space in the midbrain underlies the barn owl's skill in open-loop localisation. Certainly, the greater density of space-specific neurons devoted to the 30° arc in front of the face would account for the owl's greater accuracy in locating sounds in this region. The large proportion of the neuronal map that is devoted to the region of space somewhat below and in front of the face also makes good functional sense because this is likely to be the prey-containing region for an owl scanning the ground from its perch.

Barn owls normally use both sight and hearing to track their prey, and in keeping with this a combined auditory and visual map of space is found in the optic tectum. The auditory map in the tectum is derived by topographical projection from that in the auditory midbrain and is precisely aligned with the visual map of space derived from the retina. The mutual alignment is adjusted through sensory experience early in the life of the owl (Knudsen, 1983, 1998). In turn, this sensory map in the optic tectum is linked via neural circuits in the hindbrain to the motor circuits responsible for generating head movements (Masino & Knudsen, 1990).

6.4 Synthesising a neuronal map of auditory space

Each space-specific neuron in the owl's midbrain responds to stimuli delivered via the earphones only when both the time difference and the intensity difference fall within the range to which it is tuned. It is not excited by either the correct time difference alone or the correct intensity difference alone. Evidently, the receptive fields are formed by tuning of the neurons to specific combinations of time differences and intensity differences, which are coded separately by lower-order neurons. The initial separation takes place at the level of the earliest staging post of the auditory pathway in the brain, consisting of the two cochlear nuclei, the angular nucleus and the magnocellular nucleus.

The information available to the brain is coded in spikes generated by the sensory neurons of the owl's inner ear. Each of these auditory neurons responds to a particular frequency of sound, and the neurons produce their spikes at or near a particular point on the arriving sound wave. This latter property is called phase locking and it is important for measuring a sound's time of arrival accurately. The number of spikes generated on each occasion is proportional to the sound pressure level at the ear. Thus, the train of spikes travelling along each axon of the auditory nerve carries information about both the time of arrival and the intensity of a particular sound frequency. When it reaches the brain, each of these sensory axons divides into two branches; one enters the angular nucleus and the other enters the magnocellular nucleus.

The role played by these two nuclei is revealed by an ingenious experiment in which a tiny amount of local anaesthetic is injected into one of them so as to inactivate most of its neurons (Takahashi, Moiseff & Konishi, 1984). The responses of the space-specific neurons are then re-examined to see what changes have occurred. The results are clear cut: inactivation of the angular nucleus alters the responses of the space-specific neurons to interaural intensity differences without affecting their responses to time differences, and inactivation of the magnocellular nucleus alters the neurons responses to time differences without affecting their responses to intensity differences. When recordings are made from the interneurons in these two nuclei, their properties are found to be consistent with this result. Neurons of the magnocellular nucleus preserve the phase locking shown by the sensory neurons but are insensitive to changes in intensity, whereas the neurons in the angular nucleus are sensitive to intensity changes but do not show phase locking. Evidently, these two cochlear nuclei serve as neural filters, which pass along information about either time of arrival or intensity, but not both.

The first place in the auditory pathway at which information from both ears is compared is the lamina nucleus, which receives excitatory input from both the left and right magnocellular nuclei. It seems clear from the anatomy and physiology of the lamina nucleus that this staging post serves to measure interaural time differences. The laminar neurons are arranged

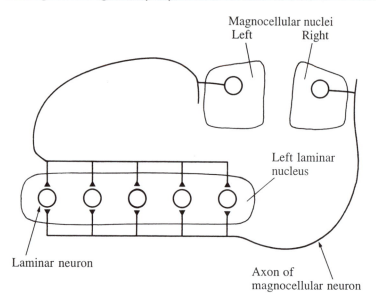

Figure 6.5 A neuronal circuit for measuring interaural time differences in the barn owl, shown as a highly diagrammatic section through the brain. The left laminar nucleus receives excitatory input from both the left and right magnocellular nuclei, represented here by a single axon from each. When spikes conducted along the left and right magnocellular axons reach a given laminar neuron simultaneously, that neuron will be strongly excited. This will happen whenever the difference in a sound's arrival time at the two ears compensates for the difference in time taken for spikes to travel to that laminar neuron along the left and right magnocellular axons. (Modified after Konishi, 1992.)

in an elongated array, and the axons of the ipsilateral magnocellular neurons pass along this array from one end whereas the contralateral magnocellular axons pass along from the other end (Fig. 6.5). Both ipsilateral and contralateral axons give off terminal branches that make synaptic contact with the laminar neurons.

This circuit is able to compute interaural time differences on the principle of **delay lines** and coincidence detection. The laminar neurons fire maximally when they receive excitatory input simultaneously from both ipsilateral and contralateral magnocellular neurons and so function as coincidence detectors. The magnocellular axons function as delay lines because the time it takes each spike to travel along the axon from one end

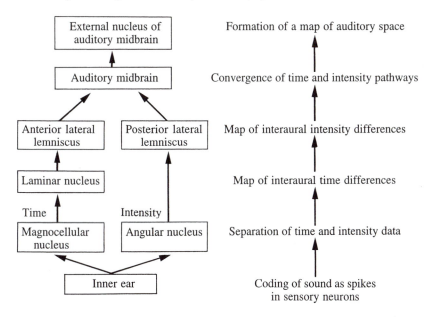

Figure 6.6 A simplified flow diagram showing how the neuronal map of auditory space is synthesised in the brain of the barn owl. The boxes on the left represent successive regions of the brain, and the process that takes place in each region is shown on the right of the corresponding box. The arrows indicate the flow of information along pathways (left) and along the sequence of computational steps (right). Note the separation of the time and intensity pathways. (Modified after Konishi, 1992, 1993.)

of the lamina nucleus towards the other end causes a delay in the spike's time of arrival at a given laminar neuron. The point in the array of laminar neurons where the left and right magnocellular spikes arrive simultaneously will, therefore, vary systematically as a function of the difference in a sound's time of arrival at the left and right ears plus the conduction time along the magnocellular axons from the two ears. Consequently, for each frequency band, the array of laminar neurons effectively constitutes a neuronal map of interaural time differences (Carr & Konishi, 1990). The axons of the laminar neurons convey this information forward to the auditory midbrain, via the anterior part of the lateral lemniscal nucleus (Fig. 6.6).

The posterior part of the lateral lemniscus on each side of the brain receives axons directly from the contralateral angular nucleus. This input is excitatory and is in proportion to the sound pressure level at the contralateral ear. At the same time, each lateral lemniscus receives inhibitory input from the lemniscal nucleus on the opposite side of the brain and this reflects the sound pressure level at the ipsilateral ear. So the response of the lemniscal neurons depends on the balance between the degree of excitation and the strength of inhibition, which in turn depends on the intensity difference between the ears. Furthermore, the neurons of the posterior lateral lemniscus vary systematically in the intensity difference that causes them to respond maximally, forming a topographical array along the dorsoventral axis (Manley, Koppl & Konishi, 1988). In effect, this array constitutes a neuronal map of interaural intensity differences for a given frequency band.

This information is passed along by the axons of the lemniscal neurons to the auditory midbrain, where it is eventually combined with the information on interaural time differences to generate the receptive fields of the space-specific neurons (Fig. 6.6). By this means, the neuronal map of auditory space is synthesised centrally from sensory cues that are not themselves spatially organised. The inner ear is organised tonotopically, with the topographical array of receptors coding frequency rather than space. For each frequency band, the ear monitors intensity and time of arrival, cues that are separated in the brain and processed in parallel pathways. In both pathways, the information from the left and right ears is brought together to form a topographical array of interaural differences. Finally, the intensity and time pathways are brought together, thereby enabling neurons in the auditory midbrain to encode specific combinations of time and intensity differences.

6.5 The echolocation sounds of bats

Whereas owls locate prey by listening passively to the noises produced by the prey, insectivorous bats actively interrogate the environment using the technique of echolocation (Fig. 6.7). For this purpose, a flying bat produces a succession of loud calls, each of which consists of a brief pulse of sound. These sounds are often described as ultrasonic because they contain frequencies (20–200 kHz) beyond the range of human hearing. The sound pulses travel out in front of the bat and, when they encounter a target, are reflected back and picked up by the ears. The essence of echolocation lies

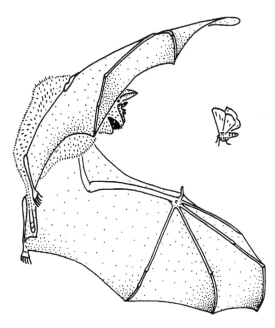

Figure 6.7 A greater horseshoe bat (*Rhinolophus ferrumequinum*) about to capture a moth in flight. The complex structure between the ears and the mouth is the nose-leaf, through which echolocation sounds are emitted in horseshoe bats. Insects are usually captured by being scooped up in the wing membrane rather than being seized in the mouth. (After a photograph by S. Dalton.)

in the ability of the brain to reconstruct features of the target, most importantly its position in space, by comparing the neural representation of the echo with that of the original signal.

Analysis of the sound pulses used for echolocation reveals what is at first sight a bewildering variety of form as one compares one species with another, but it has become clear that there are basically only two kinds of sound signal used by bats. The first kind are broadband signals, which consist of short pulses, less than 5 ms in duration, that are frequency modulated (FM). An example of this kind of signal is found in the mouse-eared bat (*Myotis*) in the widespread family Vespertilionidae: each pulse starts at a high frequency and sweeps downwards in frequency during the course of the pulse (Fig. 6.8*a*, *b*). At any given instant, the sound within the pulse is a fairly pure tone, corresponding to the fundamental frequency generated by the larynx, with traces of a second harmonic towards the end of the pulse.

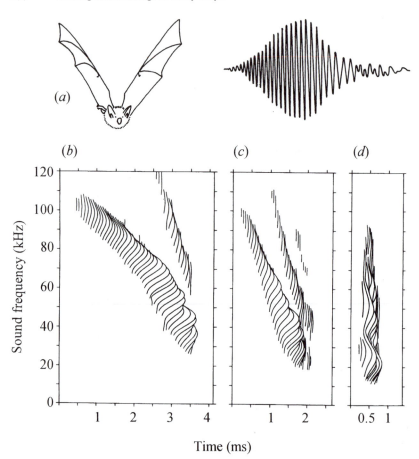

Time (ms)

Figure 6.8 Echolocation sounds of the mouse-eared bat (*Myotis myotis*). (*a*) A single frequency-modulated pulse, recorded from a bat in flight (left) and displayed on an oscilloscope (right). (*b*) Computer-generated sonagram of a single pulse, showing the downward sweep in frequency in more detail. The computer generates the sonagram by plotting curves of the relative intensity of different frequencies at successive intervals of time during the sound pulse. This pulse was emitted by the bat at a distance of 4 m from the target; (*c*) and (*d*) are sonagrams of pulses emitted respectively at 36 cm and at 7 cm from the target. (*a* modified after Sales & Pye, 1974; *b–d* from Habersetzer & Vogler, 1983.)

But the downward sweep results in the pulse having a total bandwidth of some 60–70 kHz.

Frequency-modulated pulses appear to have evolved in bat echolocation because the wide range of frequencies makes them suitable for target

description and accurate ranging. Behavioural tests show that vespertili-
onid bats can discriminate between targets that differ only in distance
(range) with an accuracy of 10–15 mm. When these bats are tested with a
pair of loudspeakers, each producing an electronically synthesised 'echo'
after each echolocation pulse, they can discriminate differences between
the speakers down to about 60 μs, which corresponds to a difference in
target range of about 10 mm. Hence, it seems probable that bats estimate
target range from the time it takes sound pulses to travel out to the target
and return as echoes, just as radar sets do. This is confirmed by studies of
the brain, which show that bats take advantage of FM signals for target
ranging by making multiple estimates of pulse–echo delay with interneurons tuned to different frequencies within the FM sweep (see section 6.8).

The second basic kind of echolocation sound used by bats consists of
narrow band signals, in which the sound has a constant frequency (CF).
These signals are generally longer, between 10 and 100 ms in duration, and
form part of an alternative strategy of echolocation employed by many
species. This alternative is particularly well developed in the horseshoe
bats, which are members of the specialised family Rhinolophidae (see Fig.
6.7). The echolocation sound of the greater horseshoe bat (*Rhinolophus ferrumequinum*) consists mainly of a long component (about 60 ms) with a
constant frequency of just over 80 kHz, which is followed by a brief downward frequency-modulated sweep and is often preceded by an even briefer
upward sweep (Fig. 6.9).

A long CF signal is unsuitable for target description but is well suited to
measuring the Doppler shift, which is the shift in sound frequency experienced by an observer listening to a moving sound source such as a passing
train. That horseshoe bats actually perceive the Doppler shifts generated
during echolocation is shown clearly by the way they modify their sounds
in flight. If a horseshoe bat is trained to fly down a long room to a landing
platform, it is observed to alter the sound frequency of its echolocation
pulses so as to keep the Doppler-shifted echoes from the landing platform
at a constant, species-specific frequency, which is 83 kHz for the greater
horseshoe bat.

In human affairs, measurement of the Doppler shift in a radar signal
provides an accurate estimate of the relative velocity of a moving target;
radar measurement of motorists' speed is perhaps the most familiar
example. Similarly, bats using a long CF signal are able to determine the

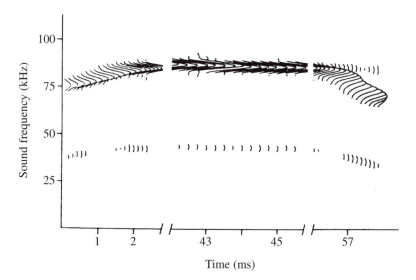

Figure 6.9 Echolocation sound of the greater horseshoe bat (*Rhinolophus ferrumequinum*). The computer-generated sonagram shows the three components of the long pulse: the initial, upward sweep in frequency; part of the long constant-frequency component (note breaks in time axis); and the final, downward sweep in frequency. The faint traces of the fundamental frequency at around 40 kHz indicate that the broadcast frequency is actually the second harmonic. (From Neuweiler, Bruns & Schuller, 1980.)

relative velocity of their prey by comparing the frequency of the outgoing pulse with that of the returning echo. In addition, it has been found that horseshoe bats are able to perceive the relatively small Doppler shifts in echo frequency produced by the beating wings of a flying insect. They use this as a means of detecting insect prey in the face of the extensive echo clutter produced by dense foliage or other background objects (see section 6.9).

Constant frequency and frequency-modulated sound pulses thus represent two different strategies for extracting information from the environment by means of echolocation. It is evident that *Rhinolophus* depends largely on the former, whereas *Myotis* employs the latter exclusively. However, there are a number of other genera that employ a mixture of the two strategies and emit pulses in which both CF and FM components are well developed. One example that has been the subject of a detailed neuroethological study is the moustached bat (*Pteronotus*) in the family Mormoopidae.

6.6 Interception of flying prey by bats

The way in which bats use their echolocation signals to track prey has been studied by combining high-speed photography with tape recordings of the normal sequence of airborne interception. These observations are necessarily made under controlled conditions but they have been supplemented by numerous, and increasingly exact, observations made in the field. All the species studied so far follow a fairly standard routine, which can be broadly divided into three stages, known as the search, approach and terminal stages. Much the same sequence is followed whether the bat is intercepting prey, avoiding an obstacle or landing on a perch.

During the search stage, a bat emits sound pulses of a constant, species-specific form at a low repetition rate of about 10 Hz or less, as described in the previous section. As its name implies, the main function of the search stage is to detect potential prey or obstacles. Behavioural tests with bats using FM signals show that they can detect a small sphere (0.5 cm diameter) at almost 3 m and a larger sphere (2 cm diameter) at over 5 m. Calculation of the intensity of echoes returning from targets at these distances suggests that the maximum range of echolocation corresponds with the threshold of hearing in bats.

However, in natural interception, bats do not visibly react to targets at these distances nor do they appear to react to progressively larger targets at progressively greater distances. Instead, the onset of the approach stage, which represents the first visible reaction of the bat to the target, occurs when the bat is between 1 and 2 m away in nearly all cases. Therefore, it is probable that the approach stage begins at a critical distance between the bat and the target and does not necessarily begin when the target is first detected. The transition to the approach stage is marked by the bat turning its head, especially its ears, directly towards the target and by an increase in the repetition rate of the echolocation sounds to a value of about 40 Hz. In bats such as *Myotis*, which use only FM pulses, the pulses become shorter but the slope of the FM sweep becomes steeper so that the bandwidth of the signal is maintained (see Fig. 6.8c). Species that use long CF pulses for Doppler shift echolocation do not drop the CF component during the search stage but it becomes shorter and the small FM component increases in bandwidth.

There is thus a shift towards brief, FM pulses at the approach stage in all species of echolocating bats. It is probable that bats are taking advantage of the greater information content of broadband signals as they approach the target, especially because the decision about whether or not to catch an item of potential prey is evidently made during the approach stage. When a mixture of living insects and similar-sized plastic discs or spheres is thrown into the air for bats to catch, the bats break off pursuit of the plastic objects at the end of the approach stage. Again, bats trained to discriminate between two targets in flight make their choice, as judged by the orientation of their ears, towards the end of the approach stage. The increase in repetition rate of the echolocation sounds is also understandable as a response to the need for increasingly frequent estimates of range and direction as the bat closes upon the target. When the approach stage goes to completion, it normally takes the bat to within 50 cm of the target.

The transition to the terminal stage is marked by an abrupt increase in pulse repetition rate, which rises to about 100 Hz or even 200 Hz in some cases. This increased rate clearly provides a rapid updating of information about the target's position as the bat makes its final manoeuvres to capture the target. In most species, the pulses emitted during the terminal stage are FM sweeps, often with several harmonics, that are 0.5 ms or less in duration (see Fig. 6.8d). Only in horseshoe bats, and others that exploit the Doppler shift, is a CF component retained during the terminal stage; even then, the CF component is reduced to a length of about 10 ms or less.

Bats do not usually capture flying insects in their mouths but rather use their outspread wing or tail membranes as a scoop for collecting the prey. The accuracy with which this is accomplished is illustrated by experiments in which horseshoe bats are fed on flour-covered mealworms thrown into the air. Flour marks are then found on the wing membrane at the base of the third to fifth digits, in an area with a diameter of 2 to 3 cm (cf. Fig. 6.7). This suggests that the bats are consistently able to localise prey in space to within about 1 cm^3.

6.7 The auditory system and echolocation

When echolocation sounds return as echoes to a bat, they are received by an auditory system that conforms to the general mammalian pattern. The sounds are collected by the external ear and enter the ear canal, where they impinge on the tympanum (Fig. 6.10). The vibrations of the tympanum are

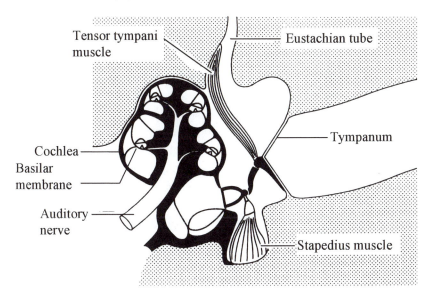

Figure 6.10 The middle and inner ear of a bat; diagram based on a horizontal section. The external ear opening is off picture to the right. Structural elements of the middle and inner ear are shown in solid black; the surrounding bone of the skull is shown in stipple. Note that the bones of the middle ear are acted upon by various muscles, known collectively as the middle ear muscles. (Redrawn after Sales & Pye, 1974.)

transmitted by the bones of the middle ear to the oval window of the inner ear. From here, the vibrations travel through the cochlea along the basilar membrane, which forms a helical ribbon, wide at the apex of the cochlea and narrow at its base. Transduction takes place in receptor cells, the hair cells, which are distributed along the length of the basilar membrane. The receptor potentials that are produced in the hair cells by the vibration of the basilar membrane are transmitted across synapses to first-order neurons with axons that carry spikes to the brain along the auditory nerve.

The information that is provided by the traffic of spikes in the auditory nerve is processed through a sequence of levels in the mammalian brain, as it is in other vertebrates. The auditory pathway consists of a rather complex network of brain regions but may be illustrated by three main staging posts. Firstly, there is the cochlear nucleus in the hindbrain, which receives input from the auditory nerve and contains mainly second-order neurons. Secondly, there is the inferior colliculus, which is one of a number of important nuclei in the midbrain containing higher-order auditory neurons.

Thirdly, there is the auditory cortex in the forebrain, which is the final staging post of the auditory pathway. Echolocation is possible because a bat's auditory system has several striking specialisations that enable it to receive and analyse faint echoes. These specialisations start at the peripheral level, as described below.

The first of these specialisations is that a bat's hearing is particularly sensitive to sounds that have similar frequencies to its own echolocation pulses. This is shown by testing bats with sounds of different frequencies and measuring the auditory threshold, which is the lowest sound intensity that elicits a detectable response. The results are expressed as threshold curves, in which the sound pressure level at threshold is plotted against frequency (Fig. 6.11). In bat species that use broadband (FM) signals in the search stage, it is found that the frequencies with the lowest threshold coincide with the dominant frequencies in the echolocation signal (Fig. 6.11b). Apart from this feature, the threshold curve is not very different from that of a non-echolocating fruit bat (Fig. 6.11a). Nor is the absolute threshold of hearing exceptional by mammalian standards. In bat species that use narrow band (CF) signals in the search stage, the threshold curve is much more sharply tuned to a narrow frequency band. In *Rhinolophus*, for example, hearing is very sharply tuned (Fig. 6.11c) and the echoes are kept close to this best frequency by the Doppler-shift compensation.

A second specialisation of the peripheral auditory system is that echolocating bats have highly directional hearing, in contrast to most mammals, which have good all-round hearing. This is shown clearly by behavioural tests carried out on restrained horseshoe bats. A loudspeaker directly in front of the bat's head produces 'echoes' with an electronically shifted frequency following each of the bat's CF pulses, and the bat then responds by compensating for this apparent Doppler shift. The directionality of hearing is tested by presenting sounds at the normal echo frequency from another speaker at different angles around the bat's head. When this second sound is perceived by the bat, it effectively masks the first sound and the bat does not show the compensation response. Hearing proves to be most sensitive directly in front of the head, and sensitivity falls off by about 45 dB from the midline to the side. A similar fall in sensitivity occurs at angles below the horizontal but the drop is less severe above the horizontal. Consequently, returning echoes are useful only if they fall within a narrow cone in front of the head, not more than 30° off the direction of flight (Grinnell & Schnitzler, 1977).

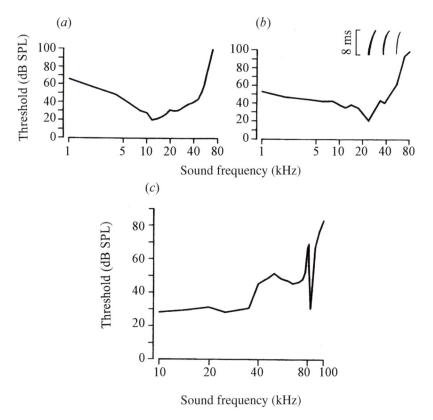

Figure 6.11 Hearing threshold curves for three species of bat, derived from recordings of summed potentials in the inferior colliculus; thresholds are expressed as sound pressure level in decibels (dB SPL), as a function of sound frequency. (*a*) Curve for a non-echolocating fruit bat from southern India. (*b*) Curve for an echolocating bat from the same locality, with a sonagram of the echolocation signal shown above for comparison (rotated through 90°). Note the similarity of the curves in (*a*) and (*b*), apart from the tuning of the curve in (*b*) to the strongest frequencies in the echolocation signal (24 to 26 kHz). (*c*) Curve for the greater horseshoe bat (*Rhinolophus ferrumequinum*) showing the sharp tuning to 83 kHz and the notch of insensitivity to frequencies just below this. (*a* and *b* modified after Neuweiler, Singh & Sripathi, 1984; *c* modified after Neuweiler, 1983.)

A third specialisation for echolocation in bats is that the peripheral auditory system shows a reduced sensitivity to the emitted pulse of sound. The echolocation sounds emitted by bats are very intense, with a sound pressure level of around 110 to 120 dB when measured 5 to 10 cm in front of the head. If the bat's auditory system were directly exposed to such intense sounds, it would not be able to recover fully by the time the echo arrived. In bats using brief FM signals, such as *Myotis*, this problem is overcome by contraction of the middle-ear muscles, which partially uncouples the inner ear from the vibrations of the tympanum (see Fig. 6.10). During echolocation, these muscles begin to contract before each sound pulse and develop maximal tension at the onset of sound emission; they then relax very rapidly, within about 8 ms, so that the response to the echo is not attenuated except at very close range. In *Myotis*, the attenuation produced by contraction of the middle-ear muscles is some 20 to 25 dB.

In addition, neural attenuation takes place in the midbrain prior to the inferior colliculus. Recording with simple wire electrodes shows that the summed potential elicited by the emitted sounds is smaller in the midbrain than in the cochlea nucleus, but this is not the case for external stimuli. Hence, there must be a central, neural mechanism that attenuates self-stimulation by the emitted sounds. This lasts only for a short time and echoes returning after 4 ms are not affected. The magnitude of the neural attenuation averages about 15 dB in *Myotis*, and so the total attenuation available from both mechanical and neural mechanisms amounts to some 35 to 40 dB.

Bats that use long CF signals, such as *Rhinolophus*, cannot exploit these mechanisms because the long outgoing pulse overlaps the returning echo. However, when the bat is on the wing, the emitted sound normally has a lower frequency than the echo, which is kept at a constant frequency by Doppler-shift compensation. Consequently, *Rhinolophus* is able to solve the problem of self-stimulation simply by being rather deaf at the normal emission frequency, which is a few kiloHertz below the echo frequency (Fig. 6.11*c*).

6.8 Auditory specialisations for echo ranging

The mechanisms of attenuation outlined above are particularly important in facilitating accurate measurement of the echo delay, and hence of distance, because they prevent the auditory system from being overloaded by the outgoing sound pulses. In order to measure echo delay, the bat's auditory system must be capable of resolving very small time intervals. Most mammals are unsuited to this task because their auditory systems take too long to recover between successive sound stimuli – usually some tens of milliseconds need to elapse after a pulse of sound before full sensitivity is recovered. But in echolocation, most echoes will return within 30 ms of the outgoing pulse, given that bats track prey at distances of no more than 3 or 4 m and that sound travels at 334 ms^{-1}.

It is therefore not surprising that the auditory neurons of bats are found to recover rapidly when stimulated with paired pulses of sound, roughly resembling the natural pairing of pulse and echo. For instance, summed potentials recorded from the midbrain of *Myotis* show some response to a second pulse that follows the first after only 0.5 ms and full recovery is achieved in 2 ms. Similar tests of hearing in fruit bats have shown that most species are slow to recover from the first pulse, but rapid recovery is found in one genus, *Rousettus*, which has evolved an echolocation capability independently of the insectivorous bats. Because this is the only respect in which the hearing of *Rousettus* differs conspicuously from that of its non-echolocating relatives, this rapid recovery may represent the most fundamental adaptation of the auditory system for echolocation (Grinnell & Hagiwara, 1972).

There is more to it than this, however, if a bat is to measure the arrival time of an echo rather than merely be aware of its existence. For echo delay to be coded by the auditory system as a time interval, some classes of inter-neurons must be able to act as accurate time markers. Interneurons that appear to be specialised as time markers for echolocation have been found in the inferior colliculus of bats belonging to several different genera. In fact, the majority of neurons in the inferior colliculus are found to have suitable properties, the most important of which is that their response is highly phasic. When one of these neurons is stimulated with a pair of FM pulses, it responds to each sound pulse by generating a single spike, or at the most two (Fig. 6.12*a*).

Repeated presentation of the stimulus shows that the delay between stimulus and response, the response latency, remains very consistent from one presentation to the next. The neuron's recovery is also sufficiently rapid that its response to the second pulse is essentially independent of that to the first pulse. As a result, the time interval between the two pulses is coded with remarkable precision in the pattern of spikes in the interneuron (Fig. 6.12*a*). Furthermore, the response latency remains almost constant regard-less of the intensity of the sound stimulus, which is a crucial specialisation

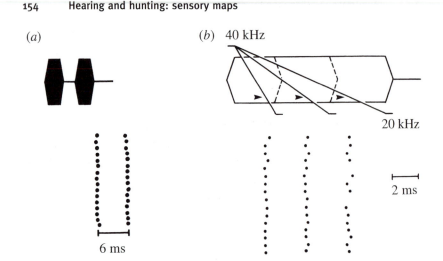

Figure 6.12 Properties of time-marking interneurons in the inferior colliculus of bats. (*a*) An oscilloscope display showing a sound stimulus, consisting
of a pair of identical, frequency-modulated pulses 6 ms apart (above), and
the interneuron's response (below). Each dot represents the occurrence of a
single spike in the responding neuron, and the pair of dot columns is generated by 16 consecutive presentations of the stimulus. (*b*) A similar display
showing the response of a collicular neuron to a single frequency-modulated pulse, consisting of a downward sweep from 40 kHz to 20 kHz,
stretched over three different durations (shown diagrammatically, above).
The interneuron's response (dot columns, below) shifts in register with the
occurrence of 25 kHz in the FM sweep (indicated by the arrowheads,
above). This suggests that the neuron is responding to this particular frequency in the stimulus. (*a* from Pollak, 1980; *b* from Bodenhamer, Pollak &
Marsh, 1979.)

for time marking in echolocation. The rate of recovery remains fast enough
for the neurons to respond consistently to the second pulse, at short pulse
intervals, even when the first pulse is as much as 30 dB louder than the
second (Pollak *et al.*, 1977; Pollak, 1980).

Each of these interneurons shows an exceptionally sharp tuning to a particular frequency within the FM sweep used for echolocation. This can be
seen in their threshold curves, which have nearly vertical slopes on either
side of the best frequency rather than the more usual V-shaped curve. As
soon as the intensity of the appropriate frequency rises a few decibels
above threshold, the neuron produces its single spike. Consequently, the
neuron is fired promptly by the same frequency component in each sound
pulse and so locks on to each event with consistent precision. This is shown

nicely by stretching the sound pulse used as a stimulus over a longer time: for each stimulus duration, the neuron responds at the same relative position in the FM sweep (Fig. 6.12*b*).

Information about time of arrival provided by these neurons is passed on up the auditory pathway, particularly to the auditory region of the cerebral cortex. Neurons that are sensitive to time delays between the outgoing pulse and the returning echo have been found in the auditory cortex of all bat species tested. Such neurons are well suited to measuring the range of a target, based on the information provided by the time-marking neurons of the inferior colliculus. It is often the case that these cortical neurons are selective for different frequencies in the pulse and in the echo. This may reflect the fact that, during active flight, the echo will inevitably be Doppler shifted to a higher frequency than the outgoing pulse. The equivalent point in time for pulse and echo will therefore be indicated by separate time-marking neurons in the inferior colliculus.

The delay-sensitive neurons of the auditory cortex respond vigorously to paired FM pulses simulating natural pulse/echo pairs but hardly respond at all to FM pulses presented singly or to CF pulses. The response consists of a short train of spikes and occurs, in the majority of neurons, only if the time interval between pulse and echo is appropriate. If one of these neurons is presented with a sequence of pulse/echo pairs, in which the echo delay is progressively reduced to simulate the bat's approach to a target, it responds strongly only to a narrow range of delays (Fig. 6.13*a*). By systematically varying the echo delay and recording the neuron's responses, it can be seen that each of these neurons is sharply tuned to a particular echo delay, termed the best delay (Fig. 6.13*b*). Some of these delay-tuned neurons even respond preferentially to a narrow range of pulse-repetition rates, which correspond to the approach stage in intercepting prey (Wong, Maekawa & Tanaka, 1992).

Neurons having similar best delays are grouped together in inwardly directed columns within a specific area of the auditory cortex (the location of the auditory cortex is shown in Fig. 6.13*c*). In the moustached bat, *Pteronotus parnellii*, these columns are arranged systematically with best delay increasing along the cortical surface from anterior to posterior (Fig. 6.13*d*). The delay-tuned neurons are thus arranged topographically according to the distance they encode and so provide a neural representation of target range. This delay-tuned area of the auditory cortex is quite distinct

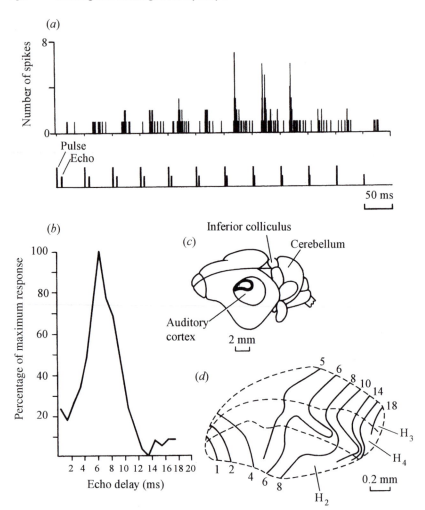

Figure 6.13 Echo-ranging neurons in the auditory cortex of bats. (*a*) A histogram (above) summarising the response of an echo-ranging neuron from
Myotis to a sequence of pulse/echo pairs (below) simulating the natural
approach to a target. (*b*) The response of another neuron from *Myotis*,
expressed as a percentage of the maximum number of spikes, as a function
of the delay between a simulated pulse and echo of constant amplitude.
(*c*) A diagram of the brain of *Myotis*, viewed from the left and above, showing
the auditory region of the cerebral cortex. (*d*) The FM area of the auditory
cortex in *Pteronotus*, showing the systematic distribution of echo-ranging
neurons according to their best delay. The solid lines labelled with a
number are contours of best delay in milliseconds. There are three horizontal clusters of neurons, indicated by broken lines, each tuned to only one of
the three harmonics (H_2, H_3, H_4) in the echo. (*a* from Wong *et al.*, 1992;
b and *c* modified after Sullivan, 1982; *d* modified after O'Neill & Suga, 1982.)

from the tonotopic area. For reasons that are not yet understood, the auditory cortex of *Myotis* is less highly differentiated: there is a tendency for neurons with larger best delays to be located more posteriorly but this hardly constitutes a neural map of best delay. Nor is the delay-tuned area clearly separated from the neighbouring tonotopic area; in fact, there is a considerable overlap (Wong & Shannon, 1988).

The echolocation pulses of *Pteronotus* contain three higher harmonics in addition to the fundamental frequency, and each delay-tuned neuron responds to only one of these three harmonics in the echo (H_2, H_3 and H_4 in Fig. 6.13d). By varying the intensity of the simulated echo when testing these neurons, it has proved possible to measure the threshold of the response at each delay tested and so to construct a threshold curve. A conspicuous result is that each of these curves has an upper threshold as well as a lower one, which means that the neuron fails to respond if the echo is too loud as well as if it is too quiet. Also, neurons tuned to shorter delays tend to have a higher threshold coupled with a narrower delay range. Each of the delay-tuned neurons is thus tuned to a particular combination of time delay and intensity appropriate to echoes returning from a certain distance (O'Neill & Suga, 1982).

The best delays in the FM area of the auditory cortex of *Pteronotus* cover a range from 0.4 ms at the anterior edge to 18 ms at the posterior edge (Fig. 6.13d), which corresponds to target ranges from 7 to 310 cm. This agrees closely with the range over which bats are observed to detect and react to targets (see section 6.6). An especially large number of neurons, reflected in cortical surface area, is devoted to delays from 3 to 8 ms (50 to 140 cm), corresponding roughly with the approach stage of target interception.

Especially with the larger values of echo delay, it is obvious that the neural response to the emitted pulse must be considerably delayed if it is to reach the cortex at the same time as the response to the echo. In fact, a large range of response latencies is found among the time-marking neurons of the inferior colliculus. It is therefore possible that the neural pathways through the colliculus are acting as delay lines and the cortical neurons are effectively coincidence detectors for particular combinations of pulse and echo latencies. The neural map of target range could thus be assembled in a similar manner to the barn owl's map of interaural time differences.

Because the FM area of the auditory cortex is sharply separated from the tonotopic area in *Pteronotus*, it is possible to inactivate them separately with drugs. As with the corresponding experiments in the barn owl's brain (see section 6.4), this provides a test of the behavioural function of these regions. When the FM area is inactivated with the drug, fine discrimination of target range is impaired, though coarse discrimination is still possible, and frequency discrimination remains unaffected. This confirms that the FM area is involved in the perception of distance, as expected from the responses of its interneurons and their topographical arrangement, but it appears to have little to do with frequency discrimination (Riquimaroux, Gaioni & Suga, 1991). The opposite result is obtained when the tonotopic area is inactivated with the drug, and this area appears to be specialised for analysis of the Doppler-shifted CF signal.

6.9 Auditory specialisations for Doppler shift analysis

All echolocating bats need to be able to analyse sound frequencies accurately. With FM signals, fine frequency analysis yields accurate range measurement and detailed description of the target, and with CF signals it yields an accurate measure of the Doppler shift. Frequency analysis is certainly excellent in bats that use only FM signals, such as *Myotis*, but their abilities are not so very different from those of other mammals. However, in bats that use long CF signals, such as *Rhinolophus*, frequency resolution is quite exceptional and greatly exceeds the abilities of non-echolocating mammals.

Specialisations for fine-frequency analysis begin at the inner ear, with the basilar membrane, which is the accessory structure that couples the sound stimulus to the receptor cells (hair cells) in the cochlea (see Fig. 6.10). Mammalian hair cells are tuned to particular frequencies according to their position on the basilar membrane, typically with equal lengths of the membrane being devoted to each octave. But in the greater horseshoe bat, the representation of frequencies from 80 to 86 kHz is greatly expanded, and the greatest expansion is found at the reference frequency of 83 kHz. This expanded representation on the basilar membrane is reflected in the number of first-order neurons that innervate the hair cells. Compared to frequencies below 70 kHz, which are not involved in echolocation, frequencies of the expanded region are over-represented approximately 10 times, with the result that 21 per cent of all first-order auditory neurons represent the frequency range from 80 to 86 kHz (Fig. 6.14*a*). By analogy with the visual system, this expanded representation is termed an acoustic fovea.

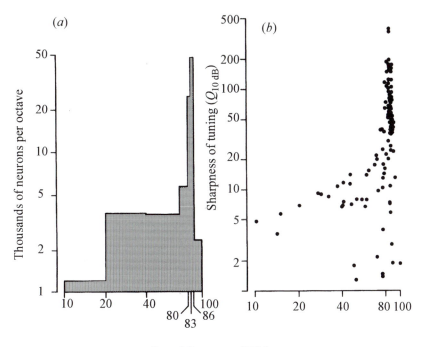

Sound frequency (kHz)

Figure 6.14 The acoustic fovea in the greater horseshoe bat (*Rhinolophus ferrumequinum*). (*a*) A histogram of frequency representation among first-order auditory neurons, showing the great over-representation of frequencies around 83 kHz. (*b*) The sharpness of tuning, expressed as the $Q_{10\,dB}$ value, for neurons with different best frequencies in the cochlear nucleus, showing the exceptional sharpness of tuning in neurons with best frequencies around 83 kHz. See also Fig. 6.11c. (*a* redrawn after Bruns & Schmieszek, 1980; *b* redrawn after Suga, Neuweiler & Moller, 1976.)

The fovea is almost certainly related to structural specialisations in the basal part of the basilar membrane, where the highest frequencies are represented. These structural peculiarities abruptly disappear at a distance of 4.5 mm from the oval window, and the expanded frequency region is located immediately beyond this critical point. As a result, the basilar membrane acts as a mechanical filter, which tunes a disproportionate length of the membrane to a narrow frequency band (Vater, Feng & Betz, 1985). A consequence of this arrangement is that the first-order neurons innervating the hair cells within the foveal region are each extremely sharply tuned. A measure of the sharpness of tuning is provided by the $Q_{10\,dB}$ value, which is the neuron's best frequency divided by the bandwidth of its threshold curve 10 dB above the minimum threshold. Very high $Q_{10\ dB}$ values of between 50 and 200 are found in the foveal region and, in the region of greatest expansion around 83 kHz, even values of over 400 are found (Fig. 6.14b). For frequencies below 70 kHz, the $Q_{10\ dB}$ values fall below 20, which is within the range found in other mammals.

This over-representation and sharp tuning are conserved at all higher levels of the auditory pathway. During echolocation, Doppler-shift compensation serves to clamp the echo of the CF component within this expanded frequency range. In effect, this behavioural response creates a constant carrier frequency, on which the small frequency modulations produced by the wing beats of flying insects are superimposed, so enabling them to be analysed by the sharply tuned neurons. A similar combination of peripheral tuning and Doppler-shift compensation has evolved independently of the horseshoe bats in the moustached bat and this combination therefore probably represents a general strategy for Doppler shift analysis.

The kinds of echo modulation produced by flying insects have been examined by the simple expedient of echolocating them with a simulated horseshoe bat, consisting of a loudspeaker broadcasting a pure tone of 80 kHz and a microphone. These tests show that the fluttering wings of an insect produce a strong echo or acoustic glint only when they are approximately perpendicular to the impinging sound waves, which happens for a short moment in each wing-beat cycle, but there are no acoustic glints from non-flying insects. A glint consists of a momentary increase in echo amplitude and a concomitant broadening of echo frequency, which represents Doppler shifts caused by the movement of the wings with respect to the sound source.

In echoes returning from a flying insect, glints modulate the echo at a rate corresponding to the wing-beat frequency of the insect, and this is termed the modulation frequency. The extent of frequency modulation involved is termed the modulation depth and is generally between 1 and 2 kHz above or below the carrier frequency. By perceiving these glints, a horseshoe bat should be able to distinguish with certainty between echoes from a fluttering insect and echoes from inanimate objects. Neurons specialised to encode the acoustic glints produced by flying insects are found among those processing the CF echo (foveal) frequencies in all the major staging posts of the auditory pathway.

At the level of the inferior colliculus, the over-representation of the CF echo frequencies is actually enhanced, being approximately 24 times that for frequencies below 70 kHz. Consequently, the normal tonotopic arrangement is distorted by the substantial block of interneurons devoted to the CF echo frequency range. Within this block, the neurons are very sharply tuned and the great majority are extremely sensitive to small frequency modulations. When stimulated with sinusoidal frequency modulations that sweep as little as ± 10 Hz around the 83 kHz carrier frequency, these neurons respond with discharges that are phase-locked to the frequency modulation. The response remains phase-locked at all modulation frequencies up to about 500 Hz, but some neurons show a preference for rates between 20 and 100 Hz. The ability of these collicular neurons to follow the complex modulations produced by real wing beats is confirmed by using the recorded echoes from a flying moth as stimuli in the experiments. With this natural stimulus, the neurons reliably encode the wing-beat frequency of the moth, as well as more subtle features of the echo (Pollak & Schuller, 1981; Schuller, 1984).

The encoded information is passed on to the tonotopic area of the auditory cortex, within which neurons devoted to the CF echo frequencies form a disproportionately large block, often termed the CF area. Neurons in the CF area respond to sinusoidal frequency modulations in much the same way as the collicular neurons but are more selective in the range of modulation to which their response is phase-locked. Most of them prefer modulation depths around ± 1 kHz and modulation frequencies of 100 Hz or less, with a strong preference for frequencies between 40 and 70 Hz. Among the nocturnal moths that are potential prey for horseshoe bats, many species have wing-beat frequencies around 40 to 60 Hz, and such moths produce acoustical glints with a modulation depth of about 1 kHz. The phase-locking of the cortical neurons is thus tuned to a behaviourally relevant range, in contrast to the wide range of responses found in the collicular neurons.

Behavioural observations confirm that horseshoe bats do in fact detect their natural prey by means of these modulations of the CF echo. Newly caged horseshoe bats only pursue insects that are beating their wings and ignore stationary insects or those walking on the sides of the cage. The caged bats will take dead, tethered insects when these are associated with an artificial wing-beat simulator placed nearby. The crucial role of the CF signal is also shown by observations on horseshoe bats foraging in flycatcher style in natural forests: while hanging on twigs, they scan the surrounding area for flying insects and take off on a catching flight only after the prey has been detected. During this stationary scanning for prey, the bats emit long CF pulses without any FM component, indicating that prey detection is based on the CF signal alone (Link, Marimuthu & Neuweiler, 1986; Neuweiler *et al.*, 1987).

6.10 Conclusions

A hunting animal depends on adequate sensory systems in order to detect its prey and to pinpoint the prey's location in space. Bats and owls have auditory systems that are dedicated to the task of detecting and localising prey by means of sound. In both groups, specialisations are found at many levels of the auditory pathway that enable spatial information about the prey to be extracted from measurement of the intensity, frequency and time of arrival of sound at the ears. Hearing in owls is adapted to process the noises made by their prey because owls depend on listening only, but hearing in bats is adapted to process the echoes of their own cries because bats hunt and navigate by echolocation.

Bats employ two basic kinds of echolocation signal, which reflect different strategies for extracting information about the prey from the returning echoes. Frequency-modulated sounds are used for prey description and for estimating the distance to the prey, both of which are based on accurate analysis of a wide range of frequencies. Distance is estimated from the delay between sound emission and echo arrival: this time interval is encoded by sharply tuned neurons that respond promptly to a given frequency in the outgoing pulse or in the returning echo. This enables higher-order neurons to be tuned to particular echo delays and so to transform the peripheral responses into a neural map of target range.

Constant-frequency sounds are used for prey detection, whether in open spaces or among dense clutter by picking out fluttering prey from the background. Analysis of CF echoes is based on an acoustic fovea, consisting of a large number of auditory neurons that are very sharply tuned to a small range of frequencies, combined with a behaviour pattern, Doppler-shift compensation, which clamps the CF echo within this range. This combination makes the auditory system extremely sensitive to small Doppler shifts, the acoustic glints, imposed on the echo frequency by the wing beats of a flying insect.

Barn owls are able to detect the angular direction of prey from sounds made by the prey. The angle in azimuth is determined from the time of arrival of the sound at the two ears, and the sound's angle of elevation is determined from intensity differences between the ears. This cue for elevation is made possible by the specialised arrangement of non-neural accessory structures, the external ear openings and associated feathers. On the basis of input from the two ears, auditory interneurons are tuned to particular combinations of time and intensity differences that correspond to particular locations in space. These neurons are arranged to form a neural map of acoustic space, which underlies the owl's skill in open-loop localisation of the prey's sounds.

In both bats and owls, the higher levels of the auditory pathway tend to be organised functionally into discrete populations of interneurons, each specialised for encoding a small set of acoustic parameters. Each set of parameters represents a particular category of information that is directly relevant to the animal's hunting or navigational behaviour, such as echo delay and intensity representing target range. Within each of these discrete populations, the neurons are arranged topographically so that a particular value of an acoustic parameter is encoded at a particular place in the brain.

Brain areas that encode sound frequency are arranged tonotopically, an arrangement that is produced by simple topographical projection from the array of receptors in the cochlea. But the representation of acoustic space or target range is synthesised by neuronal interactions and does not arise by topographical projection of the receptor array. Discrete parallel pathways extract different types of information from the simple frequency and intensity coding carried out by the receptors. Individual neurons at the higher levels then respond only to a specific combination of parameters synthesised by neuronal interactions. Thus, they act as feature detectors for specific values of behaviourally relevant information, such as angular location or target range.

Further reading

Fenton, M.B., Racey, P. & Rayner, J.M.V., eds. (1987). *Recent Advances in the Study of Bats*. Cambridge: Cambridge University Press. This book contains some useful reviews of bat echolocation; in particular, Suthers and Wenstrup and also Roverud show how much insight has been gained from behavioural studies; O'Neill reviews target ranging in the auditory system; and Vater compares the highly specialised mechanisms of frequency tuning in *Pteronotus* and *Rhinolophus*.

Konishi, M. (1993). Listening with two ears. *Sci Amer* **268**(4), 34–41. This article explains how the neuronal map of auditory space in the barn owl is synthesised from separate pathways processing time and intensity data from the ears.

Neuweiler, G. (1990). Auditory adaptations for prey capture in echolocating bats. *Physiol Rev* **70**, 615–41. This looks at the echolocation signals of bats and their corresponding auditory specialisations, in the context of the bats' natural habitat.

Popper, A.N. & Fay, R.R. eds. (1995). *Hearing by Bats*. New York: Springer-Verlag. Contained in this book are detailed reviews of major topics in bat echolocation. The overview by Grinnell and the review of the auditory cortex by O'Neill are the most relevant to the material covered in this chapter.

Suga, N. (1990). Biosonar and neural computation in bats. *Sci Amer* **262**(6), 34–41. The article deals almost exclusively with *Pteronotus*, describing its echolocation sounds and showing how both the FM and CF components are represented in the auditory cortex.

7 Programs for movement

7.1 Introduction

Understanding the mechanisms which generate and control locomotory movements is fundamental to a complete knowledge of the neuronal control of behaviour. We can regard locomotion, such as jumping, walking or flying, as basic building blocks for much of an animal's behavioural repertoire; and we can pose three basic questions about the control of such movements. First, what mechanisms ensure that muscles contract in the appropriate sequence? In walking, for example, the basic pattern is repeated flexion and then extension of each leg, with flexion of the left leg coinciding with extension of the right. Second, how does a nervous system select, initiate and terminate a particular type of movement? For example, what initiates the pattern of walking; and how is walking rather than running or swimming selected? Third, how is the basic pattern for movement modulated appropriately? Stride pattern changes, for example, when a person walks up a flight of steps or turns a corner.

Experimental approaches to these questions have often involved work on invertebrates and lower vertebrates, animals in which the parts of the nervous system that generate programs for movement contain a limited number of neurons. This offers the opportunity to identify and characterise all the components involved in generating a particular movement. A specific question that has occupied many investigators is how to determine the source of rhythmical activity that underlies many regularly repeated movements, such as walking or flying. One possible source is **proprioceptive reflexes**, and muscles could be activated in a particular sequence by joining reflexes in a chain. This view originated from Sherrington's elucidation of spinal reflexes in mammals and predominated for the first half of the twentieth century. However, a number of experiments, particularly on insects from 1960 onwards, showed that proprioceptive reflexes are not required for the co-ordination of quite long, complex sequences of movements. From these experiments emerged the concept that the central nervous system contains **central pattern generators**, responsible for generating programs for movement.

Nowadays, it is widely acknowledged that timing cues for movements are generated both within the central nervous system and by feedback from proprioceptors. Because more than one mechanism operates, the overall control of a rhythmical movement is rugged. Even within the central nervous system, it now appears to be usual that central pattern generators employ several mechanisms that operate in parallel to generate rhythms of activity. One mechanism is by circuits, such as pairs of neurons that inhibit each other so that excitation alternates between the two neurons. Another way is by **pacemaker neurons**, which have the intrinsic property of generating regular, clock-like waves of depolarisation and repolarisation of their membrane potentials.

A recent development is the realisation that circuits of neurons responsible for generating movement patterns are often quite plastic. In the locust flight system, for example, such plasticity allows the animal to adapt its movements in the best way for maintaining its course. Another aspect of plasticity is found when neurons participate in more than one type of movement. This has been most extensively investigated in a small group of neurons responsible for controlling movements of the teeth and foregut of spiny lobsters and crabs, which have shown how neuronal circuitry can be dramatically and rapidly reconfigured.

7.2 Locusts and their flight

A number of basic concepts of the way in which rhythmical movements are generated have arisen from experiments on locust flight. Locusts are infamous for their swarms, which can destroy enormous areas of crops in tropical and subtropical regions. Study of locusts in the laboratory was first promoted by the establishment of the Anti-Locust Research Centre in London, and locusts have become a favoured animal in research for many groups of neurobiologists (Burrows, 1996). Locusts are large insects, and easy to maintain and breed in the laboratory. Their nervous system is amenable to analysis at the single cell level. Locusts are capable of long migratory flights, covering several hundred kilometres a day at heights greater than 1000 m. In the laboratory, they can generate sustained flight-like movements even when securely tethered to a bar, or dissected to allow access to the nervous system.

7.3 The flight engine

As in most insects, flying is achieved by cyclical movements of four wings, borne on the posterior two segments of the thorax. As the wings move up and down, they twist so that they constantly generate lift to keep the insect airborne (Fig. 7.1a). Each wing is moved by ten muscles, which can be divided into three main groups (Fig. 7.1b). The first group contains just one large muscle for each wing, oriented along the animal's long axis. When these dorsal longitudinal muscles contract, they distort the stiff cuticular box structure of the thorax in a manner that causes the wing tips to move downwards, and they are called 'indirect depressor' muscles. The other two groups of muscles lie upright in the thorax and pull directly on the wing base. One group pulls outside the fulcrum of the wing hinge, and so these muscles are direct **depressors**, and the other group pulls on the inside of the fulcrum and so these muscles are direct **elevators** (Fig. 7.1c). The three largest direct depressor muscles of each wing control the way the wing twists around its long axis, and are important in altering the pattern of wing beat during steering manoeuvres.

The pattern of innervation of the wing muscles is straightforward because each flight muscle is usually controlled by just one or two motor neurons. A spike in a flight motor neuron mediates a rapid, strong twitch of its muscle. The motor neurons make synapses along the length of each muscle fibre, and these operate in the same way as synapses in the central nervous system. When the presynaptic terminals are depolarised by the arrival of a spike, each releases a tiny squirt of neurotransmitter, in this case the amino acid glutamic acid. The ion channels in the muscle cell membranes that bind glutamic acid cause large EPSPs. Through a series of events, the electrical signal is transduced into the development of tension by individual muscle fibres.

The simple pattern of innervation of locust flight muscles contrasts with that found in most vertebrate skeletal muscles, where each muscle is controlled by several tens or hundreds of motor neurons. It also

(*a*)

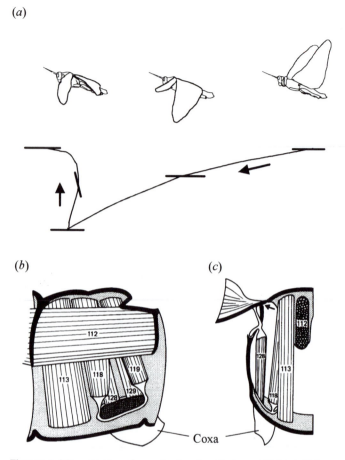

(*b*)   (*c*)

Coxa

Figure 7.1 Movements and muscles involved in locust flight. (*a*) Movements during one wing-beat cycle. The drawings show three stages during a downstroke; the hindwings move slightly before the forewings. Below, the path of movement of the tip of the left forewing during one wing beat and the angle of the wing during one wing-beat cycle are shown. The wing is twisted during the upstroke to help maintain lift throughout the cycle. (*b*) and (*c*) The main flight muscles of the third thoracic segment (bold outline in the left drawing in *a*). The right half of the segment is shown in medial view (*b*) and anterior view (*c*). Muscles 112, 128 and 129 are wing depressors; muscles 113, 118 and 119 are wing elevators. The wing hinge is indicated by an arrow in (*c*). (*a* redrawn after Pringle, 1975; *b* and *c* redrawn after Snodgrass, 1935.)

contrasts with the innervation of other muscles in locusts, such as those which move the legs. Each leg muscle is innervated by only a few motor neurons, but not all of them mediate fast, twitch-like contractions. Some, the slow motor neurons, mediate small, individual twitches that sum together gradually to develop graded, strong, slow contractions when the motor neuron generates a train of spikes. Other motor neurons are inhibitory and employ gamma aminobutyric acid as their neurotransmitter. The inhibitory motor neurons oppose the action of the slow excitors by causing hyperpolarising postsynaptic potentials in their muscle fibres. For further information about muscle innervation, see Aidley (1998).

7.4 The flight program

Recordings of the electrical signals that cause contractions of a muscle are called **electromyograms** (EMGs). The technique employed for recording EMGs during locust flight is similar to that used for recording electrocardiograms in humans, except that the electrodes are fine wires, inserted through small holes in the cuticle and secured in place with glue or wax (Fig. 7.2a). It is possible to fit a locust with a tiny radio transmitter to record EMGs from pairs of muscles while it is flying unrestrained (Kutsch *et al.*, 1993). More usually, a locust is tethered, often to a solid bar but sometimes to a harness attached to counterweights so that some of the forces generated by the wing movements can be measured. It is quite easy to induce a suspended, tethered locust to flap its wings in a flight-like manner by blowing a current of air over its head. The air current excites wind-sensitive sensory hairs. During straight, level flight the wing-beat frequency is between 15 and 20/s. The frequency tends to fall gradually during a long flight, perhaps because the locust needs less power after it has burned some of its fuel.

Each flight motor neuron usually produces one or two spikes per wing beat, and excitation of elevator motor neurons alternates with excitation of depressors (Fig. 7.2b). Over a range of different wing-beat frequencies, the delay between excitation of the elevator and depressor motor neurons is fairly constant (Fig. 7.2c), as is the duration of the depressor activity. The most variable event during a wing-beat cycle is the duration of the burst of spikes in elevator motor neurons (Fig. 7.2c). Hindwing depressors spike

(a)

(c)

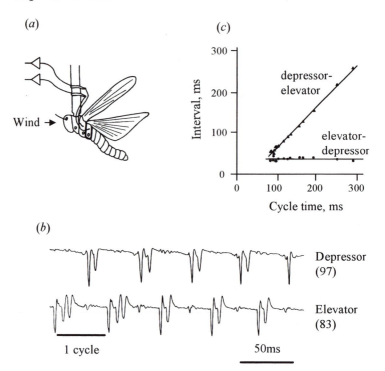

(b)

Figure 7.2 The motor program for locust flight. (*a*) The locust is tethered to a rod, and induced to fly by wind directed at the head. Fine wire electrodes are inserted into flight muscles, and are attached to amplifiers (triangle symbols). The positions of the leg attachments are shown as circles. (*b*) Electromyograms from a depressor and an elevator muscle of a forewing. Usually, two motor neuron spikes are registered in each muscle per wingbeat cycle. In midcycle, small spikes are picked up from other muscles as cross talk. (*c*) During a long flight, the gradual increase in the duration of wing-beat cycles was due to an increase in the delay between spikes in depressor and elevator motor neurons. The delay between spikes in elevator and depressor motor neurons remained constant. (*a* modified after Horsman, Heinzel & Wendler, 1983; *b* from Pearson & Wolf, 1987; copyright © 1987 Springer-Verlag; *c* modified after Hedwig & Pearson, 1984.)

5ms before forewing depressors, although elevators of all four wings are active synchronously. The result is that the hindwings are depressed before the forewings in each wing-beat cycle, so that they are not moving into the turbulent air caused by forewing movements.

7.5 Generation of the flight rhythm

An influential paper in the study of the neuronal control of movement was by Donald Wilson, published in 1960. This challenged the prevailing view that sequential activation of reflex loops generated rhythmical movements, such as locust flight and walking in vertebrates. Locust wings bear a large number of sense organs that report details of their movements. However, Wilson showed that when he removed the wings and destroyed the sense organs of the wing bases, a locust could still generate a pattern similar to flight. This consisted of regularly repeating, alternating spikes in wing elevator and wing depressor motor neurons. Either an air current directed at the head or a series of randomly timed electrical shocks to the connective nerves could elicit this flight-like activity. Flights are short in duration, and the wing-beat frequency is about half that of an intact locust. Wilson concluded that the basic program for generating flight movements is situated in the central nervous system. Generation of the rhythm, or the co-ordinated activity of different motor neurons in the correct sequence, does not require proprioceptors to report details of the movements caused by muscle contraction. Even if all the flight muscles and the head are removed, the thoracic ganglia are capable of generating a slow, repeating pattern of alternating excitation of elevator and depressor motor neurons.

Some of the neurons involved in generating the flight rhythm have been identified and characterised by intracellular recording. The thorax is opened dorsally to allow access to the thoracic ganglia, which are stabilised against movements by supporting them with a metal platform (Fig. 7.3a). Further stability is achieved by removing the legs and cutting most of the nerves that supply the flight muscles. The nerve to one muscle, usually one of the dorsal longitudinal muscles, is left intact so that EMGs recorded from this muscle provide a monitor of the flight rhythm. This nerve also carries the axons of most of the proprioceptors of the wing base. A locust prepared in this way will produce sequences of flight-like activity when wind flows over its head, or if the neurohormone octopamine is applied to the thoracic ganglia. There are many differences in the detail of the pattern from that generated by intact animals, and flight sequences are rather short in duration. However, the basic pattern of repeated, alternating activation of depressors and elevator motor neurons is present. As in an intact animal, there is a constant delay between activity in elevator and depressor motor neurons, and a delay of 5 ms between

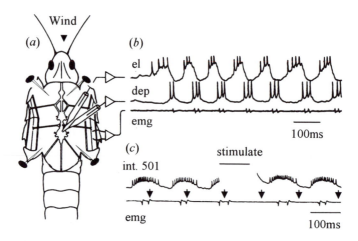

Figure 7.3 Intracellular recordings during fictive flight. (*a*) A method for preparing the locust, which is opened mid-dorsally and the body walls pinned, exposing the ganglia. Two glass capillary microelectrodes penetrate neurons in the ganglia while wire electrodes record EMGs from muscle 112. (*b*) Intracellular recordings from an elevator (el) and a depressor (dep) motor neuron, together with an EMG recording from muscle 112 (lower trace). (*c*) A phase-resetting experiment with interneuron 501. The upper trace is an intracellular recording from the neuron and the lower trace is the EMG from muscle 112 (when the neuron was stimulated with depolarising current, the intracellular recording could not be registered). The arrowheads indicate the times when bursts of spikes in the depressor muscle would have been expected to occur without any stimulus to the interneuron. (*a* and *b* modified after Robertson & Pearson, 1982; *c* modified after Robertson & Pearson, 1983.)

spikes in motor neurons of the hindwing and forewing depressor muscles. Flight-like activity in a dissected, immobile locust is called **fictive flight**.

At the start of a flight sequence, elevator motor neurons depolarise and generate spikes first. In an intact animal, this would open the wings. During flight, repeated smooth oscillations in membrane potential, up to 25 mV in amplitude, are recorded from flight motor neurons (Fig. 7.3*b*). Depolarisation of the elevator and depressor motor neurons alternates and sometimes a motor neuron might hyperpolarise from its resting potential between cycles of depolarisation.

7.6 Interneurons of the flight generator

None of the motor neurons contributes to the generation of the flight rhythm, and one good functional reason is that motor neurons, and therefore muscles, can be used independently of each other. This is important during steering, and during some movements of the legs in which muscles that also move the wings are involved. The regular waves of depolarisation that occur in motor neurons during flight must, therefore, originate in interneurons and be communicated to the motor neurons by synaptic transmission. Mel Robertson, Keir Pearson and others have characterised many different interneurons which show cycles of rhythmical activity, similar to those in motor neurons, during fictive flight. All of these interneurons generate spikes, and the interneurons generally produce a greater number of spikes per wing-beat cycle than the motor neurons. At the end of an experiment, stain is injected into a neuron so that its structure can be examined. In the thorax, about 85 different types of interneuron that are active during flight have been distinguished. All of the interneurons exist as bilateral pairs. Some probably belong to groups in which individual members have not been distinguished from each other, so that the total number of interneurons involved in the flight rhythm probably exceeds 100. The interneurons that form the central pattern generator for flight are distributed among several ganglia, and most of them branch extensively in both the second and third thoracic ganglia. In male crickets, which have very similar flight machinery to locusts, the forewing muscles are used for singing as well as for flying, and different interneurons are responsible for the two motor programs (Box 7.1).

There are two criteria for establishing whether an interneuron is part of the central pattern generator. First, it must be rhythmically active at the flight frequency; and, second, injection of a brief pulse of current into it should reset the flight rhythm. If an interneuron plays a role in generating the flight rhythm, a brief pulse of depolarising current injected into it will reset the rhythm by delaying or advancing the time of subsequent bursts of spikes (Fig. 7.3c), not only in the interneuron but also in flight motor neurons. Whether a pulse of depolarising current delivered to an interneuron advances or delays the time of the next wing-beat cycle depends on the phase of the cycle in which the stimulus current is delivered. Experiments of this type are called **phase-resetting** experiments. They reveal whether or not a particular neuron is involved in the clock mechanism that determines the timing of cycles in a rhythmically repeating activity. However, they do not reveal the exact mechanism for generating the rhythm.

One interneuron, number 301, is shown in Fig.7.4a. This neuron has its

Box 7.1. The control of singing in crickets

Male crickets sing by slightly elevating their forewings (*a*), and repeatedly rubbing a comb-like file on one against a hardened scraper on the other, shown by the cross-section through a cricket's thorax in (*b*). The pattern of activity in some wing muscles is very similar to their activity during flying. In *Teleogryllus*, during singing the wings move initially at 20 Hz, and later speed up to 35 Hz. Muscles that elevate the wings in flight cause wing closure and a pulse of sound during singing; and muscles that depress the wings in flight cause opening during singing. Despite this, different interneurons are involved in generating the two behaviour patterns. The pattern of intracellular activity recorded from motor neurons is different in the two behaviours and some interneurons are active during one activity but not the other (Hennig, 1990). This is illustrated by the recording in (*c*), which shows an interneuron producing clear rhythmical activity during singing, but not after flight was initiated by a puff of wind. This finding was surprising because it had been assumed that, because the two behaviours were similar, the same interneurons would be involved in generating the rhythms for flying and for singing. However, the evolution of singing behaviour must have involved the development of new neuronal circuitry. (*b* after Kutsch, 1969; *c* from Hennig (1990), copyright Springer-Verlag.)

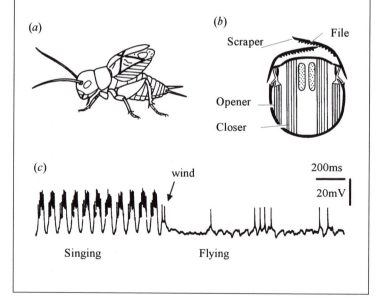

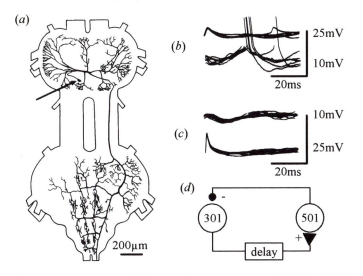

Figure 7.4 Interneurons of the central pattern generator for flight. (*a*) The anatomy of interneuron 301 in the second and third thoracic ganglia. The neuron was stained by injection of dye from an intracellular electrode. The cell body is indicated by the arrow. (*b*) Excitatory connection from interneuron 301 to interneuron 501, demonstrated by simultaneous intracellular recordings from the two neurons. Multiple sweeps of the oscilloscope are overlain, each triggered by a spike in 301. (*c*) Inhibitory connection from 501 to 301. (*d*) Schematic circuit representing the excitatory (+) and inhibitory (−) connections between 301 and 501. (Modified after Robertson & Pearson, 1985.)

cell body in the second thoracic ganglion, and many branches both in this ganglion and the third thoracic ganglion. Another interneuron, number 501, is arranged the other way round, with its cell body in the third thoracic ganglion. Intracellular recording from 501 shows that this interneuron is excited at the same time as depressor motor neurons in each wing-beat cycle. Interneuron 301 is excited slightly before the activation of depressor motor neurons.

When intracellular recordings from interneurons 301 and 501 were examined in detail, interactions between the two interneurons were found. Consistency in these interactions is shown by overlaying several sweeps on the oscilloscope. Each sweep is triggered from a spike in one of the interneurons. A spike in 301 is always followed, after a delay of 6 ms, by a small depolarising potential in 501; and a spike in 501 is always followed, after 3 ms, by a brief, hyperpolarising IPSP in 301 (Fig. 7.4*b, c*). Allowing for time for neuronal signals to be conducted to the recording sites, the delay of only 3 ms in transmission from 501 to 301 suggests that this connection is direct, or monosynaptic. The greater delay in the connection from 301 to 501 leaves room for at least one additional neuron to be involved.

These two neurons form part of a circuit that could generate bursts of activity (Fig. 7.4d; Robertson & Pearson, 1985). Excitation of 301 leads to excitation of 501, after a short and fixed delay; and 501, in turn, inhibits 301 and terminates its burst of spikes. If 301 were excited from another source, it would become active again once 501's excitation had died away, and the circuit would reverberate on its own. There is some evidence that the circuit can work in this way because, in some experiments, steady excitation of 301 by current injection causes steady, rhythmical activity in 501 and in flight motor neurons. However, it is not feasible to perform the critical test of isolating this circuit from others in the thoracic ganglia. The circuit is just one of many that have been found in the flight pattern generator. It is unlikely that any single interneuron is indispensable for generating the flight pattern.

The mechanism by which 301 excites 501 has not been fully elucidated. An interneuron called number 511 is known to be interposed between 301 and 501 (Fig. 7.5). Neuron 301 inhibits 511, which in turn inhibits 501. This implies that the delayed excitation which 301 causes in 501 is by way of two successive inhibitions, a kind of interaction called **disinhibition**. However, for disinhibition to work in this circuit, it is necessary for 511 to cause continuous, tonic inhibition of 501. This would mean that its membrane potential was continually depolarised above the threshold for its synapses to release transmitter. Only in this way can a discrete inhibitory potential in 511, caused by a spike in 301, be converted into a discrete depolarising potential in 501. Physiological characteristics of the depolarising potential in 501 suggest that it is caused by a reduction in inhibition rather than by a conventional excitatory synapse. However, we do not know whether output synapses from 511 release transmitter tonically, without requiring spikes.

The circuit in Fig. 7.5 represents a small part of the central pattern generator for flight, and allows glimpses into its mode of operation. Wind on the head activates pathways that excite 206. Interneuron 206 excites elevator motor neurons through 504, and causes delayed excitation of depressor motor neurons through 301. As explained above, a feature of the flight program is that the duration of bursts of spikes in depressor motor neurons

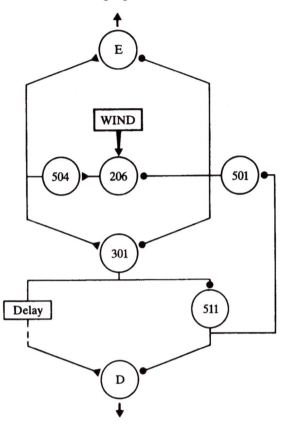

Figure 7.5 Synaptic relationships of some of the elements of the central pattern generator for flight in the locust. Excitatory connections are indicated as triangles and inhibitory connections as circles. Interneurons are numbered; and E and D indicate elevator and depressor wing motor neurons. (Modified after Robertson & Pearson, 1985.)

is relatively constant over a range of different wing-beat frequencies. This constancy can be explained by the actions of 301 in indirectly exciting these motor neurons while at the same time removing inhibition of them from 511. This creates a discrete time slot, following a burst of spikes in 301, during which it is possible to excite the depressor motor neurons. Another feature of the flight program, the constant delay between bursts in elevator and depressor motor neurons, can be explained by the dual action of 504 which excites the elevator motor neurons directly and the depressor motor neurons indirectly through interneuron 301.

Not all of the thoracic interneurons that are involved in flight participate in generating the rhythm. For example, some interneurons seem to play a role in initiating and maintaining flight, but not in the timing of wing beats (Pearson *et al.*, 1985). The 204 interneurons are a small group of about four on each side of the second thoracic ganglion. They all have axons that describe a tight loop in that ganglion and then travel anteriorly, so they cannot make direct contact with most of the interneurons of the flight generator, and they are excited by air currents blown at the head. Injection of depolarising current into a single 204 neuron to excite it strongly can trigger fictive flight, and injection of hyperpolarising current makes it less likely that air currents will trigger a flight. These interneurons may, therefore, be part of pathways that can start the flight motor pattern generator. During fictive flight, they can spike tonically, so they neither contribute to nor receive timing information for wing-beat cycles. Some stimuli which initiate flight, such as wind to the tail end of a locust, do not excite the 204 interneurons, which means that these neurons are not obligatory for starting the flight motor program, so that different sensory pathways can trigger the program independently.

7.7 Proprioceptors and the flight motor pattern

Locusts have a large array of sensors that report back to the central nervous system on the mechanical effects which the commands it issued slightly earlier caused. This is necessary because the effects of muscle contraction are not totally predictable. A flying locust might, for example, experience a gust of wind which causes its course to deviate. When it deviates, the wings on one side of the body might move up and down more than those on the other side for a few wing-beat cycles.

A large number of proprioceptors monitor movements of each wing and two types have been most intensively studied. The first consists of just one sensory receptor cell, called the wing hinge **stretch receptor**. Its cell body is embedded in an elastic strand which spans the joint of the wing with the side wall of the thorax. The large calibre of the axon of the stretch receptor has enabled experimenters to record from it, even during tethered flight. Wing elevation excites the stretch receptor and it fires a burst of spikes during each wing-beat cycle, with the number of spikes reporting the extent of elevation (Fig. 7.6a). The second sense organ is called the **tegula**, and this

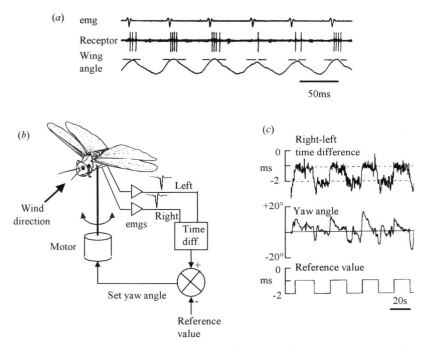

Figure 7.6 Proprioceptors and locust flight. (*a*) Spikes in a forewing stretch receptor recorded simultaneously with up and down movement of the wing and an EMG from a hindwing depressor muscle. (*b*) Arrangement of an experiment in which a locust controls its angle of flight into the wind by varying the interval between spikes in two flight motor neurons, here one on the left and the other on the right. Wire electrodes record EMGs and, after amplification, the time difference between excitation of the two motor neurons is measured for each wing-beat cycle. The time difference is compared with a predetermined reference value and if the two are equal, flight course is unaltered. If there is a difference, the motor turns by a few degrees to alter the direction of flight relative to the wind. (*c*) Recordings from an experiment like that in (*b*), in which spikes from right and left motor neurons 129 were recorded. The reference value for the right–left delay switched every 25 s between 1 and 2 ms. Quite rapidly, the locust adjusted the delay between right and left spikes to follow the reference value and, after an initial swing, it maintained a flight direction more-or-less straight into the wind. (*a* from Möhl, 1985; *b* and *c* modified from Möhl, 1988; copyright © 1985, 1988, Springer-Verlag.)

contains a number of stout hairs borne on a cuticular pad beneath the base of the wing. Inside this pad is another proprioceptive sense organ called a chordotonal organ. The hairs of the tegula are brushed and excited by wing depression. The stretch receptor and the tegula, therefore, respectively monitor wing elevation and depression. Other aspects of wing movement are also monitored; even the tiny wing veins bear tension transducers on their surfaces.

Both the stretch receptor and sensory cells of the tegula make direct, monosynaptic connections onto flight motor neurons and interneurons (Burrows, 1975; Pearson & Wolf, 1988). The stretch receptor excites almost all of the depressor motor neurons on the same side of the body as its wing. This means that when the wing is elevated, excitation of the stretch receptor will enhance excitation of the depressor motor neurons whose action opposes the elevation. If the wing is elevated more strongly than usual, the action of this proprioceptive reflex is to ensure that the depressors are excited more quickly than usual. The effect is to restore wing movement to the usual or preferred pattern. Similarly, the tegula excites wing elevator motor neurons when the wing is depressed.

In early experiments, attention was focused on the effects that stretch receptors have on the flight rhythm. When stretch receptors were removed, the flight rhythm dropped, and could be accelerated again by stimulating the axon of a stretch receptor. Recent experiments have shown that these, and other proprioceptors, play a vital role in ensuring that the flight motor output is appropriate for achieving stable locomotion. Intracellular recordings from flight motor neurons show characteristic features that are due to input from particular proprioceptors (Pearson & Wolf, 1988; Pearson & Ramirez, 1990). Elevator motor neurons, for example, show two phases of excitation during a wing beat, with synaptic inputs from the tegula sensory neurons preceding excitatory input from interneurons. The wing proprioceptors are, therefore, deeply embedded into the circuitry that generates the flight pattern.

The central pattern generator and proprioceptive feedback loops work together subtly to ensure that the animal's motor output is continually adjusted to maintain its desired course. This can be illustrated by some experiments conducted by Bernhard Möhl (1993). If a locust is tethered to a holder which allows it to swivel to the left or right, the locust will continually adjust its angle of yaw so that it flies straight into an air stream. In his experiments, Möhl did not allow the locust to provide its own power for changing its angle of yaw but used a small motor attached to the holder to twist it (Fig. 7.6*b*). Rotation of the motor was controlled by a computer, which measured the time interval between spikes in the EMGs recorded from two different flight muscles and compared this with a reference value selected by the experimenter. The locust, therefore, could control its own yaw angle by slight variations in the interval between spikes in the two motor neurons, which were often the left and right hindwing muscles 129. In the experiment shown in Fig. 7.6*c*, the first reference value chosen was a time difference of 2 ms between spikes in the left and right 129 motor neurons, so that the locust remained on a flight course straight into the wind when it maintained this interval. The reference value was then switched to 1 ms, and the locust quite rapidly reset the interval between spikes in the left and right motor neurons. Whenever the reference value was altered, the locust reset the interval between spikes in the two selected motor neurons. Different pairs of muscles were selected in a series of experiments.

These experiments have revealed a degree of plasticity in the locust nervous system which was unexpected. The locust cannot know which pair of motor neurons the experimenter has selected for a particular experiment. This means that the locust must be able to compare the results of varying the timing among different pairs of motor neurons before discovering which ones are effective in maintaining course. Cutting the pair of muscles selected destroys the ability of the locust to maintain course in this way, which shows that proprioceptive feedback about the mechanical effects of muscle contraction is required. We do not yet know which proprioceptors are involved: the stretch receptor and tegula hairs both report movements of skeletal elements rather than individual muscle contractions; and flight muscles are not known to have sense organs similar to the muscle spindles of humans embedded in them. Möhl's experiments show that small, continual variations in the motor pattern are important because they allow flexibility in the most effective pattern at maintaining course. The central nervous system does not generate an unchanging, 'perfect' motor pattern; the pattern is tuned through the action of proprioceptors. Instead, we can view one important role for proprioceptors as participating in selecting the most effective program for maintaining a desired course. The ability to make continued, small adjustments to the motor output means that the pattern can be continually updated, and rapidly adjusted to particular conditions. For a locust, these conditions would include wind direction, and turbulence caused both by the wind and neighbouring locusts in a swarm. A similar strategy is becoming adopted by engineers in designing control systems for robots, because it provides a way to compensate for inevitable inaccuracies in manufacturing. It is difficult, for example, to ensure that the friction in wheel axles on the left and right sides is exactly balanced.

7.8 Steering and initiating flight

Flying animals require sensory mechanisms both for maintaining a certain flight course and for altering it so that they avoid collisions or predators. The DCMD neuron (see section 5.8) might be responsible for the ability of locusts to avoid colliding with each other in dense swarms; and locusts, like some other insects, possess neurons which respond to the high-frequency calls of insect-eating bats and which can probably cause a locust to steer its flight path away from a hunting bat (Robert, 1989; Baader, 1991).

More is known about neurons that play a role in maintaining a particular flight course, called DN neurons. These neurons are **multimodal** – they respond to a number of different sensory modalities in a way that could report details of deviations from a straight flight course. Many of them are excited by wind flowing over the head, which is detected by groups of short, curved hairs mostly situated dorsally, between the compound eyes. Each hair is attached to a bipolar sensory neuron which projects into the brain and probably makes synapses directly with some of the large interneurons. Other sense organs which affect the interneurons include the simple eyes or ocelli, the compound eyes, and proprioceptors that monitor movements of the neck joints.

One wind-sensitive interneuron is called the tritocerebral commissure giant (TCG), after the small nerve which contains its axon. The small nerve runs between the two connective nerves near to where they leave the brain, and this unusual anatomical feature has enabled experimenters to record spikes from the TCG in loosely tethered, flying locusts (Bacon and Möhl, 1983). When a locust is flying into a steady air stream, the TCG does not spike at a steady rate. Instead, it generates a brief burst of spikes to coincide with elevation during each wing-beat cycle. The reason is that movement of the wings and nodding motions of the head create their own air currents, which interact with wind caused by the locust's movement forward through the air. This interneuron is, therefore, rhythmically active during flight. Stimulating it artificially, to induce extra spikes, can reset the flight rhythm by lengthening or shortening the cycle in which the stimulus is delivered. Therefore, the TCG meets the same criteria for being part of the central pattern generator as some of the local, thoracic interneurons.

Two additional roles have been assigned to the TCG. First, it could initiate flight during a jump, when the TCG is excited by the rush of wind over the head. Stimulating the cut axon of a TCG is an effective way of initiating fictive flight (Bicker & Pearson, 1983). Second, the left and right TCGs could stabilise flight direction with respect to wind direction. If the locust yaws so that it is no longer facing directly into the wind, excitation increases to one TCG and decreases to the other (Möhl & Bacon, 1983). Stimulating one TCG electrically can induce yaw movements during tethered flight.

Another example of a DN neuron is DNC (Fig. 7.7a). Besides being excited by wind currents, DNC responds to rotation of the visual horizon. When the horizon rolls about the long axis of the locust, DNC is excited by movements which roll the animal towards the side contralateral to its axon (Fig. 7.7b), and inhibited by movements which roll it in the opposite direction. In experiments, the thorax of the locust is fixed, and a panorama painted on the inside of a sphere can be moved around the head to simulate movements of the horizon. Inputs from the wind-sensitive and visual pathways sum, so that when a locust is flying along a straight course, DNC is tonically excited by air currents. Rolling towards one side causes an increase in spike frequency, and rolling to the other side causes a reduction in spike frequency. Probably, two separate visual pathways converge on DNC: one from the compound eyes, and the other from the simple eyes, or ocelli, which have enormous fields of view and monitor changes in horizon position by measuring changes in the total amount of light they receive (Wilson, 1978). If a locust is carefully prepared to allow access to a short length of connective nerves in the neck, but is capable of moving its wings, abdomen and head freely, steering movements can be produced by stimulating DNC with trains of electrical pulses delivered through a microelectrode in its axon (Hensler & Rowell, 1990). A train of spikes lasting a few hundred milliseconds causes a change in the relative latency of action potentials in motor neurons of steering muscles within one wing-beat cycle (Fig. 7.7c), and causes the head to roll after a slightly longer delay. Besides

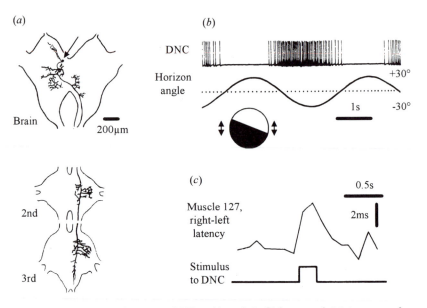

Figure 7.7 Brain neuron DNC and its role in flight control. (*a*) Anatomy of the DNC neuron with its axon on the right of the nervous system. The cell body, on the left of the brain, is indicated with an arrow. In different experiments, intracellularly injected dye revealed the structure of DNC in the brain, and in the second and third thoracic ganglia. (*b*) Spikes recorded from the axon of the right DNC in response to a wind current and to rolling motion of an artificial visual horizon. Clockwise motion of the horizon – as if the locust was rolling away from the side of the DNC axon – excited this DNC. Rolling in the opposite direction inhibited it. (*c*) Electrical stimulation of a DNC axon induced a change in the relative latency of spikes in the left and right hindwing motor neurons, 127. Similar shifts in relative latency were also recorded in response to rolling movements of the visual horizon. (Modified after Hensler & Rowell, 1990.)

the TCG and DNC neurons, between 20 and 30 other DN neurons on either side of the brain send their axons to the thoracic ganglia (Hensler, 1992). The ways in which these neurons generate steering movements are not clear, but involve some direct output connections with motor neurons as well as connections with local, thoracic interneurons.

7.9 Overall view of locust flight

The overall picture we now have of the flight control system of the locust is of a series of intermeshed, overlapping neuronal loops (Fig. 7.8). It contains

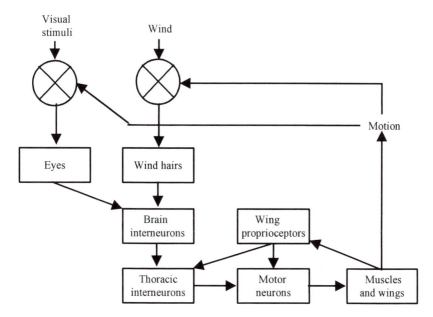

Figure 7.8 A flow diagram to show the relationships between some of the elements involved in generating the flight motor program in a locust. Some, but not all, of the thoracic interneurons included here are involved in generating the rhythm. Rhythmic input to motor neurons and interneurons is also derived from wing proprioceptors and brain neurons. As a result of the forward and rather irregular movement through the air, both the wind currents and the visual stimuli that the locust experiences are modified, so that the locust receives stimuli that are a combination of those caused by the external environment and those resulting from the locust's own movements.

a surprisingly diverse array of different nerve cells. Many of them, including the proprioceptors and at least some of the wind-sensitive interneurons, are rhythmically active at the wing-beat rhythm. We must consider these neurons to be part of the flight generator itself because they are able to reset the flight rhythm in the same way as some of the thoracic interneurons. The flight generator is highly **redundant** – it contains many elements with similar or overlapping functions, none of which is indispensable to its normal operation. This redundancy brings the advantages, first, that the flight system is rugged and not easily perturbed, and, second, that the pattern produced is flexible and able to adapt rapidly to changing demands from the environment. Flight involves additional elements, not shown in Fig. 7.8. One of these is that the neurohormone octopamine is released into the haemolymph. Octopamine affects the contractile properties of muscle cells (Box 7.2) and increases the supply of fuel to muscles. In addition, it can affect the properties of some of the interneurons of the flight motor program generator, making them liable to fire spikes in cyclical bursts (Ramirez & Pearson, 1991), although it has not been demonstrated that octopamine is released within the central nervous system during flight.

A strong stimulus for flying is wind on the locust's head and, in the laboratory, flight ceases rapidly when wind stops. However, many rhythmical movements of animals are initiated by a brief stimulus and continue for some time after the stimulus finishes. Two quite different behaviours of this kind are examined below, both of which serve to take an animal away from a noxious stimulus. The first behaviour is escape swimming by the large nudibranch mollusc *Tritonia* to escape from starfish predators. The smell of starfish elicits fairly stereotyped movement patterns in many marine invertebrates and in *Tritonia* they consist of about four to seven cycles of alternate flexion of the body upwards and then downwards. The second example is swimming by young tadpoles of the frog *Xenopus*, which is elicited by touching the body surface.

7.10 Triggering and maintaining escape swimming in *Tritonia*

Some of the first intracellular recordings from nerve cells in an animal behaving in a more-or-less normal pattern were made from *Tritonia* (Willows, Dorsett & Hoyle, 1973). By exposing the brain through a small incision and then fastening it to a support platform, intracellular recordings can be made from single neurons. The animal is suspended in a tank of seawater and can perform normal muscular movements. This experimental technique allowed identification of motor neurons and interneurons that are involved in local withdrawal reflexes as well as in the dorsal and ventral movements that occur during escape swimming (Fig. 7.9*a*). Ideas about the neuronal mechanisms that control escape swimming have been revised quite radically a number of times, and both the behaviour and the neuronal circuits that mediate it are more simple than those involved in locust flight.

A swimming episode lasts up to a minute, and consists of four to seven cycles of alternating bursts in dorsal and ventral flexion neurons. As in locust flight, interneurons generate the rhythm and communicate it to

Box 7.2. The neuromodulator octopamine

Octopamine, like some other amines, has a wide variety of effects on behaviour. In lobsters, injection of octopamine into the blood causes an animal to adopt submissive behavioural postures, whereas another amine, serotonin, causes a lobster to behave aggressively (Kravitz, 1988). There is considerable interest in elucidating the roles of these substances in behaviour: do they help prepare the body of an animal for certain kinds of behaviours, or do they also act as the triggers for particular behavioural acts? In locusts, octopamine has widespread effects. For example, it activates pathways that metabolise fat to provide energy during flying; it increases the responsiveness of proprioceptors such as the wing hinge stretch receptor; and it potentiates transmission at connections made by the stretch receptor in the central nervous system (Orchard, Ramirez & Lange, 1993). It is released from some of the dorsal unpaired midline (DUM) neurons which are thought to release octopamine from diffusely placed swellings of their axons rather than from conventional synapses. In (*a*), a DUM cell in the third thoracic ganglion of a locust was stained by intracellular injection. Sites of octopamine release, if any exist, are unknown in the central nervous system. One effect, which octopamine mediates directly on the contractile proteins, is to increase the rate of relaxation following a twitch. This is shown in recordings of twitches by the muscle that extends the hind tibia of a locust (*b*). Each twitch is caused by a spike in this muscle's slow motor neuron. Support for a role for DUM neurons in preparation for particular movements comes from the observations that some DUM neurons are active before and during flight (Ramirez & Orchard, 1990), and others are active just before a kick by the hind legs (Burrows & Pflüger, 1995). (*a* modified after Watson, 1984; *b* modified after Evans & Siegler, 1982.)

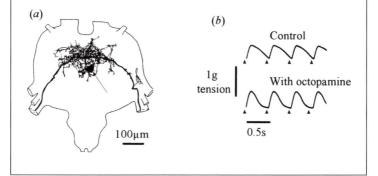

(*a*)

100μm

(*b*)

Control

1g
tension

With octopamine

0.5s

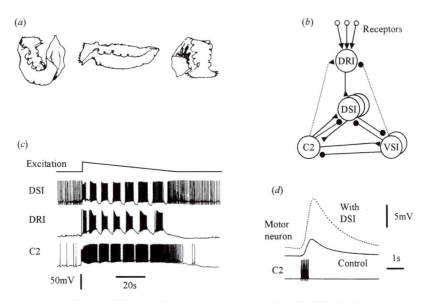

Figure 7.9 Escape swimming in the sea slug (*Tritonia*). (*a*) Swimming consists of several alternating dorsal and ventral flexion movements. These movements do little to propel the animal through the water; they lift it from the substrate, and it is carried by water currents. (*b*) The pattern of connectivity between interneurons involved in generating the program for swimming. In each side of the brain, receptors, including chemoreceptors, excite DRI, the dorsal ramp interneuron. The swim rhythm is generated by a network involving three dorsal swim interneurons, (DSIs), two ventral swim interneurons (VSIs), and one C2 interneuron. (*c*) Intracellular activity recorded simultaneously from a DSI, C2 and DRI during a swim episode, caused by electrical stimulation of a sensory nerve. At the start of the swim, the DSI is excited strongly, and its excitation declines slowly in a ramp-like manner during the swim. The source of this excitation is the DRI. (*d*) Modulation of the strength of a synapse from C2 to a motor neuron by activity in a DSI. A short burst of spikes in C2, elicited by intracellular stimulation, caused an EPSP in the motor neuron. When stimulation of C2 was preceded by stimulation of a DSI for several seconds, the size of the PSP caused by C2 increased dramatically. A similar increase was also produced by bathing the brain in seawater containing serotonin. (*b* and *c* modified after Frost & Katz, 1996; copyright National Academy of Sciences, USA; *d* from Katz *et al.*, 1994; reprinted with permission from *Nature*; copyright © 1990 Macmillan Magazines Ltd.)

motor neurons. Three groups of interneurons on each side of the brain are involved in rhythm generation: three dorsal swim interneurons, two ventral swim interneurons, and one other interneuron called C2 (Fig. 7.9*b*). None of these interneurons has an intrinsic capacity to generate rhythmical oscillations in membrane potential, and the rhythm is generated by a circuit of linked interneurons. In an isolated brain, electrical stimulation of a nerve elicits a very similar pattern of intracellular activity to that which occurs in an intact animal that is swimming in response to starfish odour (Fig. 7.9*c*).

The swim neurons are strongly excited at the start of a swim. As the swim progresses, the excitation gradually diminishes and the cycle period becomes longer. A single interneuron on each side of the brain collects input from sensory neurons and conveys the excitation to the swim interneurons (Frost & Katz, 1996). The interneuron is called the dorsal ramp interneuron (DRI) because its excitation decays in a slow, ramp-like manner during a swim. Excitation of the DRI always precedes that of other interneurons before a swim starts. During a swim, a DRI continues to spike, with a brief interruption during each ventral flexion caused by an inhibitory synapse from the ventral swim interneurons. Stimulating a DRI with enough current to make it spike at similar frequencies to those that occur during a swim will trigger swim activity in the network. There are similarities between a DRI and some of the wind-sensitive interneurons of the locust flight system. Both respond to the same sensory stimuli that trigger a particular motor pattern, and electrical stimulation of the neurons triggers the pattern. However, the DRI is unlike any single wind-sensitive interneuron in that swimming will not start unless it is excited. If hyperpolarising current is injected into a DRI so that the DRI does not spike when a peripheral nerve that normally elicits swimming is stimulated, swimming does not begin. Also, ramp excitation of the three swim interneurons does not occur. If a DRI is hyperpolarised after a swim has started, the swim episode stops. The DRI, therefore, acts as a channel through which all sensory excitation must pass in order to activate the central pattern generator for swimming.

Each DRI makes excitatory connections with the three dorsal swim interneurons, which play a dual role in generating the program for swimming (Katz, Getting & Frost, 1994). First, the dorsal swim interneurons contribute to the pattern of each cycle of the rhythm through their synaptic connections with the ventral swim interneurons and with C2. Second, they sustain the rhythm because the neurotransmitter that they release, serotonin, has multiple effects within the network. Not only does serotonin act as the neurotransmitter at the output synapses made by a dorsal swim interneuron, it also increases the excitability of C2 and enhances the release of neurotransmitter from C2. For example, the EPSPs which C2 mediates in a motor neuron increase in amplitude when a dorsal swim interneuron is excited (Fig. 7.9d). The facilitatory action which the dorsal swim interneurons exert on the output synapses from C2 is called **intrinsic neuromodulation**, because the neuromodulator is released by neurons that are intrinsic to the pattern-generating network and is only released when the pattern generator is active. The neurons of the swim generator are also involved in less dramatic, local withdrawal responses. When swimming starts, the dorsal swim interneurons generate an intense burst of spikes, and it is likely that the serotonin which this releases ensures that the network is configured to generate co-ordinated, rhythmical swimming movements.

7.11 Swimming by young *Xenopus* tadpoles

The most complex patterns of behaviour occur in vertebrates, and understanding the neuronal basis of these behaviours must involve unravelling the circuits of the spinal cord. One approach, which has been particularly fruitful, is to examine very simple behaviours in lower vertebrates, particularly swimming in lampreys (Grillner *et al.*, 1995) and in tadpoles. The pattern of swimming in newly hatched tadpoles of the clawed frog *Xenopus*, turns out to be especially simple (Roberts, 1990). For the first few days of life, these tadpoles spend much of their time attached to leaves and stems by a cement gland on the head. If the skin is touched, a tadpole will swim by side-to-side undulating movements of its trunk and tail until it finds a new attachment site (Fig. 7.10*a*). At the start of a swim, the body flexes to the left and then to the right up to 20 times per second. The rate of flexion declines during a swim. Waves of movement travel towards the tail end and propel the tadpole forwards at up to 5 cm/s.

The spinal cord of a young tadpole contains about 1000 neurons of eight types, including five types of interneuron. Alan Roberts and his colleagues have pioneered a method of making intracellular recordings from these neurons during fictive swimming. A tadpole is immobilised with the drug curare, which blocks neuromuscular transmission. When the skin is touched, the pattern of spikes in motor neuron axons is the same as in free-

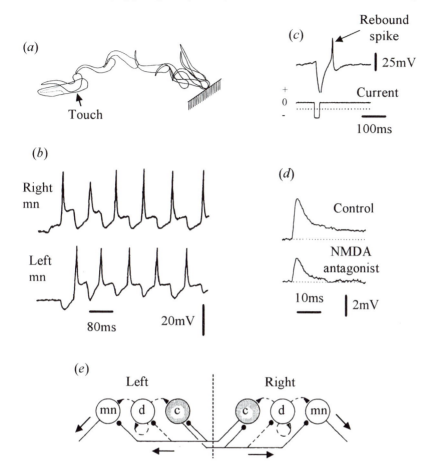

Figure 7.10 The generation of the swim pattern in *Xenopus* tadpoles. (*a*) Tracings from a video recording of a short swim by a young tadpole, initiated by touching its skin. (*b*) Intracellular recordings from motor neurons of left and right trunk muscles during fictive swimming. During the swim, both neurons were depolarised from resting potential (dotted lines). The neurons spiked out of phase with each other, and received IPSPs in midcycle. (*c*) Rebound spike in a motor neuron. A small amount of steady, depolarising current was injected into the neuron, and a rebound spike occurred at the end of a brief pulse of hyperpolarising current. (*d*) Dual component EPSP in a motor neuron, caused by stimulating the axon of an interneuron. When the spinal cord was bathed in a solution containing an antagonist of the NMDA glutamate receptor, the slow and prolonged component of the EPSP disappeared. (*e*) The network that is thought to generate the rhythm for swimming. The descending interneurons (d) are involved in circuits of mutual excitation, and the commissural interneurons (c) convey inhibitory signals across the nerve cord so that excitation of motor neurons (mn) alternates in the two sides. (After Roberts, 1990; reprinted with permission from Science and Technology Letters.)

swimming tadpoles, demonstrating that the central nervous system contains a central pattern generator for swimming. Touching the skin excites sensory neurons called Rohon Beard cells and a single spike in one of these cells can be sufficient to elicit swimming. During a swim, a motor neuron remains depolarised from resting potential. It produces just one spike per swim cycle, and receives an IPSP when neurons on the opposite side fire (Fig. 7.10b). When tadpoles mature sufficiently to feed, the pattern of activity during swimming becomes considerably more complex, and motor neurons generate bursts of spikes (Sillar, Wedderburn & Simmers, 1991).

The pattern generator for swimming provides an excellent lesson on how both synaptic circuitry and intrinsic membrane properties shape the output of a central pattern generator. Like the other rhythmical behaviours examined here, interneurons generate the pattern and, as in the case of locust flight, different mechanisms for generating the rhythm operate in parallel. An essential element of the swim pattern is that motor neurons on the left and right sides are excited alternately. This is achieved by commissural interneurons that are excited in time with motor neurons on the same side as their cell bodies and dendrites. The axons of the commissural interneurons cross the spinal cord and inhibit motor neurons and interneurons of the opposite side.

The midcycle IPSP that these interneurons mediate has a dual role. Besides ensuring that motor neurons are inhibited when their contralateral partners are excited, an IPSP triggers the next spike in the motor neuron. The spike is called a **rebound spike** because it is triggered by the end of an event that hyperpolarised the motor neuron. A pulse of hyperpolarising current injected from a microelectrode can also be a trigger for a rebound spike. However, rebound spikes are not initiated if the membrane potential simply repolarises to resting at the end of a hyperpolarising pulse. A small, continual depolarising current must also be injected, to deliver a small amount of excitation at the end of the hyperpolarising pulse (Fig. 7.10c).

During a swim, steady excitation is provided through excitatory synapses from interneurons. The steady depolarising potential sets up a critical balance between currents carried through two types of voltage-sensitive channels. The first type is sodium channels, and entry of sodium through these channels tends to excite the neuron. The second type is potassium channels, through which potassium will leave the neuron. The current carried by potassium ions will increase as the neuron depolarises, opposing the excitatory action of the sodium channels. The effect of a brief hyperpolarisation is to close both types of channel. When the hyperpolarisation ends, the sodium channels open more rapidly than the potassium channels, allowing the generation of a single spike.

The steady depolarisation is necessary for the generation of a rebound spike because it enables the sodium channels to open following an IPSP. During swimming, it is provided by a population of interneurons that excite motor neurons on their own side of the nerve cord. There are probably circuits of mutual excitation among these interneurons, which helps to sustain swimming. Another factor that sustains excitation throughout a swim is that the EPSPs which these interneurons cause in motor neurons have relatively long durations, up to 200 ms, which is much longer than one swim cycle. This is because the EPSPs have two different phases, each mediated by different types of receptor for the neurotransmitter glutamate. The two components of an EPSP can be dissected apart by pharmacological agents (Fig. 7.10*d*). The first, fast component can be mimicked by applying the drugs kainate or quisqualate to the surface of a motor neuron, while the second can be mimicked by applying another drug, *N*-methyl-*D*-aspartate (NMDA). Both of these drugs are agonists of glutamate – they bind to receptors and activate their ionic channels in a similar way to glutamate. The motor neurons, therefore, have two different types of receptor for glutamate at their input synapses: one type causes the initial, fast excitation, whereas the second type, called the NMDA receptor, mediates the longer-lasting excitation. The fast EPSPs help excite the motor neuron at the appropriate time in a swim cycle, reinforcing the rebound from an IPSP.

A characteristic of the NMDA receptor is that its channel can only open if the postsynaptic neuron is already strongly depolarised, through the action of other synapses. This feature makes the NMDA receptor suitable for responding to combinations of signals, and it is heavily implicated in mechanisms of learning. In the tadpole spinal cord, strong excitation of motor neurons by the faster-acting, glutaminergic synapses is necessary to ensure that the NMDA-mediated EPSPs can switch on.

The mechanism for generating the swim pattern, therefore, depends on circuitry that ensures that left and right sides alternate in their activity and that there is mutual excitation between interneurons (Fig. 7.10*e*). Cellular properties underlie the sustained excitation of motor neurons by NMDA receptors, and the generation of rebound spikes. Simulation by computer has shown that circuits that have these characteristics can generate an output pattern like the tadpole swim pattern (Roberts & Tunstall, 1990). However, the interneurons that convey inhibition across the cord are not essential for the generation of a rhythm because when the spinal cord is split longitudinally, each half can generate rhythmically repeating activity on its own.

Some of the features of the swim generator of young tadpoles have also been discovered in neurons of the mammalian spinal cord, including rebound excitation and long-lasting EPSPs mediated by NMDA receptors. The relative complexity of movements that the mammalian spinal cord is capable of producing might arise partly from a greater diversity of types of neuron compared with the young tadpole. However, it is equally possible that the two spinal cords contain similar circuits of neurons but, in the mammal, there are many variations in the basic wiring of these circuits. Recent work on a relatively simple ganglion, the stomatogastric ganglion of decapod crustacea, has demonstrated that circuitry can, indeed, be reconfigured extensively, allowing groups of neurons to participate in quite different activity patterns.

7.12 Circuit reconfiguration in the stomatogastric ganglion of the lobster

In crustacea, movements of the foregut are achieved by contraction of discrete striated muscles, very similar to the muscles that move limbs during movements of the body. Quite complex, regularly repeating patterns of activity occur in these muscles. A major reason for investigating their control is that the neurons responsible for generating these movements are contained in four discrete, small ganglia and the nerves that connect them (Fig. 7.11a). The most intensively investigated is the stomatogastric ganglion which, in lobsters, contains just 30 neurons. A small nerve connects it with three other ganglia of the stomatogastric nervous system. Almost all of the neurons in the stomatogastric ganglion are motor neurons and, unlike the situation for the three examples described earlier in this chapter, the motor neurons are essential components of the central pattern generator.

The stomatogastric ganglion controls two different parts of the foregut, called the gastric mill and the pyloric region. The gastric mill contains three teeth, which cut and grind food after it has been churned in a region called

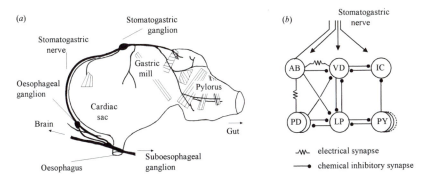

Figure 7.11 The foregut and stomatogastric nervous system of a spiny lobster (*Panulirus*). (*a*) The foregut with its ganglia and nerves. The stomatogastric ganglion lies in a blood vessel (not shown) on top of the stomach. The stomatogastric and oesophageal ganglia are unpaired, and there is a circumoesophageal ganglion attached to each of the large connective nerves which link the brain and suboesophageal ganglion. (*b*) A diagram showing some of the synaptic connections between neurons of the pyloric section of the stomatogastric ganglion. All neurons except the anterior burster (AB) are motor neurons; the two PD neurons control dilatation of the pylorus and the other motor neurons (VD, LP, IC and PY) control different phases of constriction. The stomatogastric nerve contains axons of neurons which have a variety of different effects on neurons of the stomatogastric ganglion. (*b* modified from Selverston & Miller, 1980.)

the cardiac sac. Muscles of the gastric mill are activated in a rhythm with a cycle period of 5–10 s. The pyloric region, into which the gastric mill empties, has a shorter cycle time, of 0.5–2 s; it squeezes and mixes food particles.

The pattern of synaptic connections within the ganglion and the array of active, membrane properties which single neurons possess are startlingly complex. The 14 neurons of the pyloric network are connected in a network that includes over 20 electrical synapses and over 60 chemical, inhibitory synapses (Fig. 7.11*b*). At one time it was thought that the gastric mill rhythm is primarily generated by reciprocal, inhibitory circuit interactions, whereas the pyloric rhythm is generated by intrinsic bursting in one interneuron. However, we now know that the generation of each rhythm involves both mechanisms. One experimental approach that was particularly useful in revealing the different mechanisms involved isolating single cell circuits of small numbers of cells from others in the ganglion by killing the cells that were presynaptic to the neurons of interest. To kill a particular neuron, a fluorescent dye was injected into it through a microelectrode and then an intense spot of blue light was shone onto the cell body (Selverston & Miller, 1980). Using this technique, it was possible to isolate a single pair of neurons that are connected by reciprocal, inhibitory synapses (Miller and Selverston, 1982). A simple two-neuron network is able to generate rhythmical activity, although the pattern differs from that produced by an intact stomatogastric ganglion.

Neurons of the stomatogastric nerve play a vital role in pattern generation by unmasking particular properties that are intrinsic to individual cells. The nerve contains 60–120 axons and rhythmic activity is not recorded from the stomatogastric ganglion if the stomatogastric nerve is cut or if its axons are silenced. However, rhythmic co-ordinated output from the ganglion can be restored either by stimulating axons within the nerve or by adding certain transmitters, particularly some amines or peptides, to the seawater bathing the ganglion. Because the pyloric neurons only burst in the presence of particular transmitters, they are called conditional oscillators. All of the neurons in the pyloric region are conditional oscillators but one, the anterior burster (AB in Fig. 7.11b), has the fastest rhythm and acts as the master clock, setting the pace for the whole pyloric region.

One particular bilateral pair of neurons that project into the stomatogastric ganglion, called the pyloric suppressors, have widespread effects on motor patterns (Meyrand et al., 1991; 1994). When the pyloric suppressor neurons are active, the separate pyloric and gastric mill rhythms stop; and the pyloric and gastric mill neurons become active in a new, co-ordinated pattern of rhythmic activity (Fig. 7.12a). An electric synapse links the two pyloric suppressor neurons, so they operate as a single functional unit. When the stomatogastric ganglion is expressing the pyloric and gastric mill patterns, the pyloric suppressor neurons are silent. However, they can be excited by applying food to chemoreceptors that are situated on the valve that separates the oesophagus from the stomach. When a pyloric suppressor neuron is excited, either through these sense organs or by injection of depolarising current, its membrane potential oscillates and it generates bursts of spikes. These bursts of spikes excite motor neurons which cause the valve to open, and intracellular recordings from pyloric and gastric mill neurons show profound changes in the activity patterns they express. Some neurons are tonically inhibited from producing spikes, but others, such as

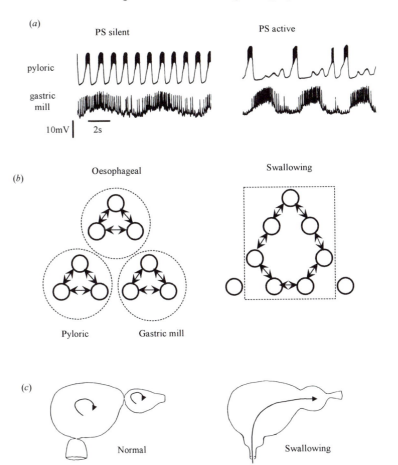

Figure 7.12 Action of the pyloric suppressor neuron (PS) in reconfiguring the stomatogastric ganglion. (*a*) Intracellular recordings from a neuron of the pyloric region and a neuron of the gastric mill region of the stomatogastric ganglion. On the left, PS was silent and the two neurons expressed different rhythms of activity. When PS was stimulated, the two neurons became active in the same new rhythm. (*b*) A diagram to illustrate the effects of PS on the circuitry of the stomatogastric ganglion. When PS is silent, networks for the oesophageal, pyloric and gastric mill rhythms operate independently of each other. Activity in PS causes some neurons of these three networks to be incorporated into a new functional network, and silences others (dotted). (*c*) Behavioural correlates of PS activity. On the left, different foregut regions are acting independently and valves between them are closed. On the right, the lobster is swallowing. (*a* from Meyrand, Simmers & Moulins,1991; reprinted from *Nature*; copyright © 1991 Macmillan Magazines Ltd; *b* and *c* redrawn after Meyrand, Simmers & Moulins1994.)

the pyloric and gastric mill neurons, express a new pattern in which neurons from both regions are co-ordinated.

The new pattern is probably that for swallowing, and the pyloric suppressor neurons also cause the valve between the oesophagus and stomach to open. During swallowing, movements of the whole foregut must be co-ordinated to move food onwards (Fig. 7.12c) and the swallowing pattern has a frequency between that of the pyloric and gastric mill regions. After the pyloric suppressor neurons have stopped firing, the new pattern persists for several tens of seconds and the original two pyloric and gastric mill rhythms are slowly re-established. This observation is important because it shows that circuits within the ganglion itself are reconfigured by activity in the pyloric suppressor neurons (Fig. 7.12b). If the stomatogastric ganglion neurons were all driven by bursts of spikes in the pyloric suppressor neurons during swallowing, we would expect the pyloric and gastric mill rhythms to be re-established as soon as the pyloric suppressor neurons stopped firing. The pyloric suppressor neurons reconfigure the circuits of the stomatogastric ganglion, so that neurons which previously participated in different and unco-ordinated activities now express a new, single pattern. Many of the details of how this reconfiguration is achieved remain to be worked out.

7.13 Conclusions

Usually, motor neurons are not involved in the generation of rhythms for movements, but are driven through synapses from interneurons and sensory neurons. This arrangement allows motor neurons to be excited independently of each other, so that their muscles can be used in different movement patterns. Networks of neurons that can generate programs for movement are often called pattern generators. In general, rhythmical movements are generated by a number of different mechanisms operating together, which makes the pattern generator robust.

Phase-resetting experiments demonstrate whether a particular neuron is involved in generating rhythmical activity. In locust flight, the rhythm is generated by networks of interneurons, reinforced by cyclical feedback from proprioceptors such as the wing hinge stretch receptors, and from wind-sensitive neurons. There is good evidence for the existence of reverberating circuits of interneurons in the flight generator, but because so many interneurons are involved it is difficult to assign a precise role to a single neuron and to disentangle different circuits from each other.

There are a number of different ways in which the excitation of a rhythm-generating network can be sustained. In locust flight, sensory activity, particularly from wind-sensitive neurons, is important. Neuromodulators can play important roles by strengthening synaptic connections, as in *Tritonia* swimming, or by switching on bursting properties, such as in the pyloric rhythm of lobsters or, perhaps, locust flight. Finally, long-lasting postsynaptic potentials such as those mediated by NMDA receptors in tadpole swim motor neurons can also be important in sustaining a rhythm.

Circuits that generate programs for movement are not 'hard wired' and inflexible. One way in which this is evident is the manner in which sensory feedback works in controlling locust flight. At its simplest, sensory feedback allows a motor program to compensate for changing demands on muscles as the nature of the terrain alters. Another more subtle role is to select between patterns that differ in their effectiveness, for example in moving an animal along a straight path. Continual small variations in motor output allow the motor program to be updated continually, allowing compensation for changes in the mechanical properties of an animal's skeleton and muscles as it grows or is injured. Interneurons can also cause radical changes in the way that networks are configured. This is well illustrated by the way in which a single interneuron can reconfigure the lobster stomatogastric ganglion so that the neurons participate in new groupings and patterns of activity.

Further reading

Arshavsky, Y.I., Orlovsky, G.N., Panchin, Y.V., Roberts, A. & Soffe, S.R. (1993). Neuronal control of swimming locomotion – analysis of the pteropod mollusk *Clione* and embryos of the amphibian *Xenopus. Trends Neurosci* 16, 227–33. A review that compares and contrasts the mechanisms that generate simple programs for swimming in frog tadpoles and in a planktonic snail. The review deals with general problems in the production of rhythmical motor activity.

Harris-Warrick, R.M., Marder, E., Selverston, A.I. & Moulins, M. eds. (1992). *Dynamic Biological Networks: the Stomatogastric Nervous System*. Cambridge, MA: MIT Press. Although this book is primarily about the crustacean stomatogastric nervous system, its reviews synthesise material in a way that throws light on the general properties of motor control systems.

Stein, P.S.G., Grillner, S., Selverston, A.I. & Stuart, D.G. eds. (1996). Neurons,

Networks and Motor Behaviour. Cambridge, MA: MIT Press. This collection of reviews was written to accompany a conference that is held every ten years to discuss and present work on the generation of motor patterns, particularly to highlight the most recent advances. The meeting is reviewed briefly in Katz, P.S. (1996). Neurons, networks and motor behaviour. *Neuron* 16, 245–53.

8 Circuits of nerve cells and behaviour

8.1 Introduction

There are very few instances in which complete neuronal pathways can be traced from the level of sense organs all the way to that of motor neurons. Notable exceptions are some startle behaviours like those described in Chapter 3, in which the size of the giant neurons involved makes experimental study relatively easy and in which the links between sensory processing and motor control are short. However, most of an animal's behavioural repertoire is not performed with the same urgency as escape movements. Much sensory analysis, particularly in visual and auditory pathways, involves several different stages, distributed over different regions of a brain. How are different sensory messages identified, and how are appropriate motor programs selected? The type of problem can be illustrated with a simple example. If a fly lands on your cheek, or if the skin of your knee itches, you can move your hand without thinking to those locations to remove the source of annoyance. This might seem like a trivial example of behaviour, but the neuronal mechanisms that allow us to perform such an act are far from being understood. It is relatively straightforward to map the locations of sensory receptors in the skin in an orderly manner within the brain, which generates a **somatotopic** map. However, it is not simple to generate the correct commands that will generate the correct balance of activation in different muscles of a jointed limb so that its end arrives at a specific location on the body surface. How is a map of position transformed into a series of finely graded muscle contractions? For an electrical engineer, it would be a major challenge to build a robot that could locate an object on its surface and then move the tip of a jointed appendage to the correct location to scratch itself.

This chapter, examines some extremely simple behaviours that involve reflex withdrawal movements of restricted regions of animals' bodies. Different experimental approaches have been adopted to study these behaviours. Microelectrodes are the most widely used tool for studying the activity of neurons, and allow connections between neurons to be characterised in detail. However, only a few neurons can be monitored at any one time by using microelectrodes. There is great interest, therefore, in techniques that allow an investigator to record the activity of many neurons simultaneously. One approach is to use optical methods, and another is to study the way that model networks of neurons work. A common theme that is emerging is that the central nervous system is rather diffusely organised, with relatively few interneurons dedicated to single behaviour patterns.

8.2 Neuronal activity during different behaviours in *Aplysia*

Aplysia is a shell-less gastropod mollusc that spends most of the time grazing on seaweed near or just below the low-tide mark. It breathes by drawing currents of water over a delicate gill, which is normally partially extended from under a fleshy shelf of the body wall, but is retracted for defence if nearby skin is touched lightly (Fig. 8.1a). The gill is also strongly retracted in time with respiratory pumping movements. Some species of *Aplysia* grow in excess of 30 cm long and have attracted the attention of neurophysiologists since the 1930s because the nervous system contains a number of neatly arranged ganglia with large, often distinctly coloured, neuron cell bodies that are ideal targets for microelectrodes. The abdominal ganglion has been particularly intensively investigated, and is the ganglion that contains cell bodies of gill-withdrawal motor neurons and of some of the sensory neurons that innervate the skin near to the gill.

Some sensory neurons make direct, monosynaptic, excitatory synapses onto gill-withdrawal motor neurons, forming direct, simple reflex pathways. These pathways are suitable for mediating the gill-withdrawal behaviour. As the intensity of a stimulus applied to the skin increases, the sensory neurons generate more spikes, and more sensory neurons are recruited. This increases the frequency with which motor neurons generate EPSPs and spikes. The classical processes of temporal and spatial summation of postsynaptic potentials in the motor neurons could readily explain the smoothly graded relationship between stimulus strength and response intensity that characterises this simple behaviour (Kandel, 1976). However,

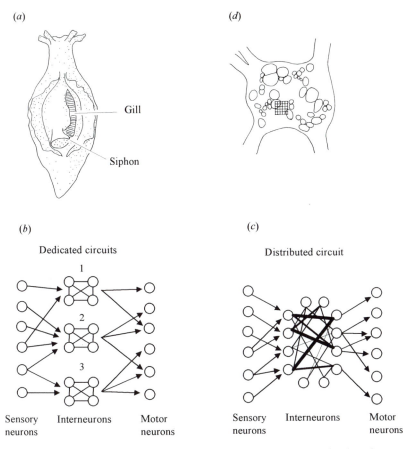

Figure 8.1 The gill-withdrawal reflex of *Aplysia*. (*a*) A view of *Aplysia* from above, showing the location of the delicate gill. (*b*) and (*c*) Two alternative ways in which circuits controlling behaviours such as gill withdrawal could be organised. (*d*) A diagram of the dorsal surface of the abdominal ganglion showing some of the larger neuron cell bodies. Superimposed is a grid of 25 squares, each of which is the size of a photocell used to monitor spikes, as explained in the text. (*b* and *c* modified after Wu *et al.*, 1994; *d* modified after Falk *et al.*, 1993.)

these direct monosynaptic pathways are not the only ones responsible for mediating the gill-withdrawal reflex (estimates of their contribution vary from 5 per cent to 60 per cent). Sensory neurons make extensive connections with motor neurons through indirect pathways that involve interneurons – this parallel arrangement of direct and indirect pathways linking sensory and motor neurons should be familiar from the way that proprioceptors influence locust flight (see section 7.7). *Aplysia* also has a peripheral nervous system. The involvement of interneurons not only increases the complexity of the integrative processes underlying the relation between stimulus and response, but also introduces the possibility that responses by motor neurons to skin stimulation will be influenced by other behaviours in which these interneurons are involved.

Besides responding to mechanical stimulation of the skin, large gill-withdrawal movements occur as part of ordinary ventilatory pumping movements, and smaller withdrawals often occur apparently spontaneously. An engineer would probably design separate circuits responsible for reflex and ventilatory withdrawal and feed the outputs of these circuits through a switch so that the gill could be controlled either by one or the other. In this kind of configuration, particular elements would be **dedicated** to specific behaviours, as shown in Fig. 8.1*b*. Circuit 1 might be responsible for generating respiratory pumping movements, while circuit 2 might organise retraction in response to touching the skin and circuit 3 might control another kind of gill retraction. All three circuits are subject to input from sensory neurons and they each drive some of the gill-withdrawal motor neurons (direct links between sensory neurons and motor neurons are omitted in the diagram). In another arrangement, drawn in Fig. 8.1*c*, interneurons are interconnected more widely and diffusely; different functions are distributed within the whole circuit, so that single interneurons participate in several different behaviours.

8.3 Optical monitoring of neuronal activity

The abdominal ganglion contains about 900 neurons, and it would be an enormous task to catalogue all their interconnections by making recordings with microelectrodes. Another approach, adopted by Lawrence Cohen and colleagues, is to monitor the activity of as many different neurons as possible while the animal performs different behaviours. To do this, Cohen *et al.* used a chemical (a pyrazolone oxonole) which has been specially developed as a tool in neurophysiological experiments. The dye attaches to cell membranes, and the amount of light that the dye absorbs alters as the voltage across the cell membrane changes. A ganglion is soaked in a solution of this dye and an image of the ganglion is projected by a microscope objective lens onto an array of 144 or 464 tiny photodiodes, each of which measures the amount of light passing through a small part of the ganglion (Fig. 8.1*d*). The electrical signal produced by each diode is amplified and processed to produce pulses that indicate the occurrence of spikes. Some neurons are large enough to cover more than one photodiode, and other diodes pick up spikes from more than one neuron. Nevertheless, these experimenters are able to process data in a way that records the occurrence of spikes in about a third of the neurons of the abdominal ganglion at one time, although they cannot assign spikes to identified neurons.

One of the first discoveries made using this technique was that hundreds of neurons in the abdominal ganglion are active during any gill-withdrawal reflex (Zečević *et al.*, 1989). This observation does not allow us to distinguish between the two arrangements shown in Figs. 8.1*b* and 8.1*c* because we need to know whether each neuron is active only during the reflex withdrawals, or during other behaviours as well. Subsequently, three different kinds of gill-withdrawal movements were studied: reflex withdrawal; weak, spontaneously occurring withdrawals; and strong withdrawals that occur during respiratory pumping (Wu, Cohen & Falk, 1994). During each of these movements, between 62 and 72 neurons were usually active and almost all of these neurons were active during all three behaviours. This means that something like 200 neurons in the abdominal ganglion, most of which are interneurons, would be excited during any activity of the gill-withdrawal muscles. The neuronal networks responsible for controlling gill-withdrawal movements seem, therefore, to be organised in a distributed pattern like that in Fig. 8.1*c*, with few neurons active during one behaviour but not during the other behaviours.

Control of gill movements is a major concern of the neurons in this ganglion and many of the other activities that it controls, such as heart beat, need to be co-ordinated with gill movements. However, the abdominal ganglion is also responsible for controlling a more dramatic defensive response to intense stimulation of the skin, which involves expulsion of a dense cloud of dark purple ink from a special gland. The inking behaviour contrasts with gill-withdrawal in that it is an all-or-none behaviour (Carew & Kandel, 1977; Byrne, 1981). The same sensory neurons that trigger gill withdrawal responses also trigger inking, and they communicate through interneurons to three, strongly coupled motor neurons. It would be interesting to investigate whether these three interneurons are dedicated to inking behaviour, or whether they are also active during other behaviours.

The abdominal ganglion of *Aplysia* is suited to optical monitoring of neuronal activity because it is quite small and transparent. However, a major drawback of this technique is that it cannot be used to stimulate or inhibit an individual neuron. In some other ganglia, the technique is capable of resolving small changes in membrane potential, so that there is a chance that it can be used to trace circuits between neurons by correlating the occurrence of spikes in some neurons with the postsynaptic potentials they cause.

8.4 Local bending reflexes in the leech

If you touch a small area of the skin of a leech, it will bend its body away from the touch, a movement caused by contraction of longitudinal muscles in a few segments close to the site of the touch. The body of the leech locally assumes a U shape. Touching the top of the body causes a downward bend, whereas touching the left side causes a bend towards the right, and so on (Fig. 8.2*a*). The main sensory neurons involved in triggering these local reflex movements are pressure-sensitive P neurons. Each segmental ganglion contains four P neurons, each of which has a receptive field that covers one quadrant of the skin of its segment – upper or lower right, and upper or lower left. The muscles that produce the bending movements run longitudinally along the body wall, and in each segment there are six bilateral pairs of dorsal, ventral and lateral longitudinal muscles. Each dorsal or ventral longitudinal muscle is controlled by two motor neurons, an excitor and an inhibitor, and each lateral muscle is controlled only by an excitatory motor neuron.

Excitation of a particular set of P neurons reliably activates motor neurons in a particular way (Fig. 8.2*b*). No direct connections between P neurons and motor neurons have been found, so interneurons must be involved in the local bending reflexes. A straightforward way of arranging the circuitry of a ganglion to control these local avoidance reflexes would be to connect sensory neurons with motor neurons through interneurons that are responsible for particular reflex behaviours, in the manner indicated in Fig. 8.2*c*. Shawn Lockery and Bill Kristan characterised these interneurons by using intracellular recording and staining (Lockery & Kristan, 1990a, 1990b). They studied interneurons that are involved in dorsal bending, the production of a downward U shape when the top surface of a leech is touched. In a survey of 73 ganglia, they found nine types of interneuron

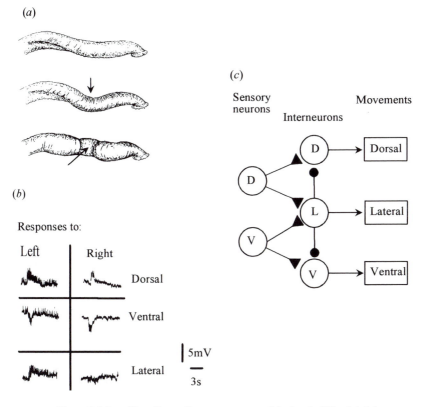

Figure 8.2 Local bending reflex movements of the leech (*Hirudo*). (*a*) Drawings of a leech viewed from its right side: resting (top), and responding to touches to the top (middle) or side (bottom), as indicated by the arrows. (*b*) Intracellular recordings from the excitor motor neuron of the left dorsal muscle in response to stimulation of P neurons. In the upper and middle recordings, the four P neurons were stimulated individually. In the lower recordings, lateral touches were mimicked by stimulating the dorsal and ventral P neurons of either side simultaneously. (*c*) A circuit that incorporates interneurons which are dedicated to particular bending movements. The triangles indicate excitatory synapses and the circles indicate inhibitory synapses. (*a* redrawn after Lockery *et al.*, 1989; *b* and *c* redrawn from Lockery & Kristan, 1990a.)

that were both activated by dorsal P neurons and made connections onward to dorsal bend motor neurons. Each type had a distinct morphology, allowing it to be distinguished from others and to have its cell body position identified on a map of a segmental ganglion. One, neuron 125, is illustrated in Fig. 8.3*a*.

Each interneuron responded to touch anywhere on the skin of its

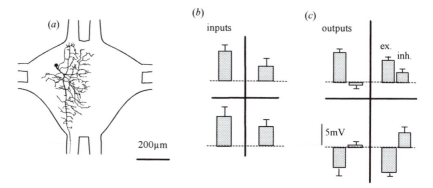

Figure 8.3 Interneuron 125, which is involved in local reflex bending movements in the leech. (*a*) A diagram of a ganglion in which the neuron was
stained by intracellular injection of the fluorescent stain lucifer yellow. (*b*)
Amplitudes of EPSPs in a left 125 caused by stimulating the four different P
neurons (dorsal and ventral left and right) of the same segment. Each bar
indicates, from three experiments, the mean amplitude and standard error
of the EPSP evoked by stimulating a particular P neuron. (*c*) Amplitudes of
postsynaptic potentials evoked by spikes in interneuron 125 in the excitor
(ex.) and inhibitor (inh.) motor neurons of dorsal and ventral longitudinal
muscles. Upward-directed bars indicate EPSPs and downward-directed bars
indicate IPSPs. Means and standard errors are indicated, as in (*b*). (*a*
redrawn after Lockery & Kristan, 1990b.)

segment (Fig. 8.3*b*), and made connections with most of the longitudinal
motor neurons in its ganglion (Fig. 8.3*c*). None of the interneurons
appeared to be dedicated to just the dorsal reflex bending behaviour
because each was also active during ventral or lateral bending. Therefore,
the networks responsible for analysing touch stimuli and triggering
bending movements seem to be arranged in a distributed manner in which
particular interneurons are not assigned to specific local bending
responses. However, is it possible that other interneurons, not found by
Lockery and Kristan, are the ones that are primarily responsible for specify-
ing particular bending responses?

One test for the involvement of a neuron in a behaviour is to remove
that neuron and observe any change in behaviour. When hyperpolarising
currents were injected into single interneurons to reduce their excitability, responses by motor neurons to stimulation of P neurons were
reduced in amplitude. But these experiments were inconclusive because
the reductions in response by motor neurons were small, which is not

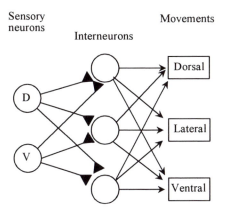

Sensory
neurons
                    Interneurons
                                        Movements

Figure 8.4 A diagram indicating the kind of organisation of the circuit that
controls local bending movements in the leech. The diagram is simplified in
many ways so that it does not indicate connections between left and right
neurons or the relative strengths of different connections, and includes only
three of the 17 interneurons known to exist in each ganglion.

surprising because each motor neuron is driven by several different
interneurons.

8.5 Modelling a network of neurons

In order to find out whether specific behavioural responses could be produced using interneurons with such unspecific characteristics, computing
techniques were used to construct a model of the network involved in leech
local bending (Lockery *et al.*, 1989). Software represented three levels of
units, with one level equivalent to sensory neurons, one equivalent to
motor neurons, and the third interposed between them like the interneurons (Fig. 8.4). The 18 interneuron units connected only with sensory and
motor neurons and not with each other, which is the same pattern as in the
leech. The strengths of connections between interneurons and sensory or
motor neurons were variable.

In an experiment, a model network was gradually trained to reproduce
the behaviour of a leech segmental ganglion, so that activation of a particular sensory neuron would generate an appropriate local bending-avoid-
ance response. The way in which this was done was first to programme the
computer to make connections between sensory neurons and interneurons and between interneurons and motor neurons. The pattern of the

connections was set randomly, except that interneurons were arranged as left–right pairs, and all the connections were initially weak. The performance of the circuit was then tested to find out how closely its behaviour matched that of the leech. It would be very unlikely to perform well in this first test. Next, the computer altered the strengths of some of the connections, and the behaviour of the circuit was again tested. If the performance of the circuit had been improved, the same connections were again strengthened, but if it had deteriorated, these connections were weakened and others were strengthened. By repeatedly altering connection strengths and then testing the circuit, the computer gradually improved the circuit in a stepwise manner until the input–output relationship became the same as that observed in a leech. This technique of training a model network of neural elements is called **back propagation**. Successful training of the network required between 10 000 and a 100 000 repetitions of the training cycle.

Eighteen different experiments were performed. Each one produced a network that reproduced the behaviour of a leech but was different from the other 17 networks. All the networks contained model interneurons that resembled the leech interneurons in the lack of specificity in their connections, with more than 95 per cent receiving inputs from two or more P cells and 88 per cent making output connections to at least seven of the eight motor neurons. This means that it is possible to adopt a strategy of using diffusely connected elements in a network to link inputs and outputs in specific ways. Indeed, because the computer program always constructed a diffuse network rather than one in which individual units were dedicated to restricted functions, the strategy must be a good one.

In a leech segmental ganglion, therefore, each of the interneurons involved in dorsal bending makes many connections that are inappropriate to that response. Stimulation of a particular area of skin always causes the same bending response, which means that the particular balance of the strengths of connections within the ganglion must offset the apparently inappropriate nature of many of them. Although computer modelling generated a number of different circuits for controlling local bending, it is likely that the network of neurons in one leech is very similar to that in another. It would be interesting to know what factors in evolution have selected this particular network. Careful use of computer modelling techniques, such as this study on the leech, is an invaluable tool for neuroethology because it allows an experimenter to test whether a neuronal network operates in the way expected. An alternative approach to reconstructing circuits is to isolate individual neurons and allow them to make synaptic connections in culture (Box 8.1).

Box 8.1. Reconstructing circuits of live neurons

A very direct test of whether a circuit of neurons is sufficient to gener-ate an activity is to isolate the circuit concerned from the nervous system. In the lobster stomatogastric ganglion, which contains a very small number of neurons this has been done by killing some neurons, so that only those involved in a particular circuit are left. Another technique was adopted by Syed, Bulloch & Lukowiak (1990) in a study of the pond snail, *Lymnaea*. This is a diving, air-breathing animal; it has a lung, and comes to the surface to exchange the air it contains by rhythmically opening and closing the aperture, called the pneu-mostome (*a*). Experiments using intracellular recording revealed a simple circuit of three interconnected interneurons that are active during the ventilatory rhythm (*b*). Is this circuit sufficient to generate the ventilatory rhythm, or are additional neurons required? In order to test this, cell bodies of the three neurons were carefully dissected from the central nervous system and placed in a dish containing tissue culture medium. Quite quickly, the isolated cell bodies began to sprout new processes and, within a day, the three neurons established synaptic connections with each other, in the same pattern as in the intact brain. Brief electrical excitation of one of the neurons caused several cycles of rhythmical activity very similar to that found in an intact brain. None of the neurons was capable of generating rhythmi-cal activity on its own and, although I.P3.1 and V.D4 (which connect with motor neurons) made a simple circuit in which each inhibited the other, rhythmical activity was not produced unless the giant, dopamine-containing cell (Rpe.D1) was also present.

(*a*)   (*b*)

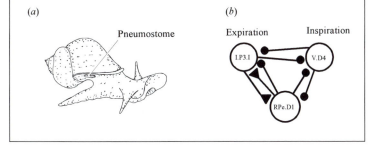

8.6 Local reflex movements of a locust's leg

Jointed limbs, such those of a mammal or an insect, are used in a wide variety of ways. Besides being involved in locomotion, they are capable of finely controlled movements such as grooming, in which the end of the limb is brought precisely to a particular spot on the animal's surface. Since the 1970s, Malcolm Burrows and colleagues have investigated in detail the neurons responsible for controlling local movements of a locust's hind leg, providing a very complete description of the links between sensory analysis and motor control (see Burrows, 1996). Compared with the local bending movements of the leech, local reflexes of a locust's leg involve many more neurons, arranged in two layers between the sensory and motor neurons. Each of these layers consists of **local interneurons**, which have their branches restricted to the third thoracic ganglion, often to one part of it. The first layer mostly receives inputs from sensory neurons and consists of **spiking interneurons**. The second layer makes connections with motor neurons, and consists of **non-spiking interneurons**. The major route for the flow of information is from sensory neurons to spiking interneurons, then to non-spiking interneurons and finally to motor neurons. Within this route there are no feedback pathways, so motor neurons do not synapse back onto either type of interneuron, and non-spiking interneurons do not synapse back onto local spiking interneurons. Feedback is provided by sense organs that monitor movements of the limbs, and these mainly connect with interneurons rather than directly with motor neurons.

If a locust's leg is touched anywhere on its surface, it will be moved away from the touch (Fig. 8.5a). Usually, the movement involves several joints so that, for example, when the ventral surface of the tibia is touched, the tibia is raised and during this response the angle between the tibia and the foot changes to keep the foot parallel with the ground. These movements are controlled by neurons contained within the third thoracic ganglion, including about 70 motor neurons and a few hundred local interneurons. About 10 000 sensory neurons project into each half of the third thoracic ganglion, and many of these originate on the surface of the leg, so there is a great deal of convergence as information flows from sensory neurons to motor neurons.

Many of the sensory neurons belong to basiconic sensilla, each of which consists of a short tactile hair with one mechanosensory neuron and four chemosensory neurons. Trichoid sensilla are another type of sense organ

(a)

(b)

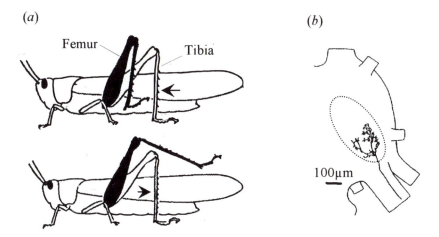

Figure 8.5 Local reflex movements of a hind leg of a locust, *Schistocerca*. (*a*) Movements evoked by touching the tibia on its anterior or posterior (large arrows). The leg is shown as an outline in its initial position and in black in its final position. (*b*) The projection of a trichoid sensillum on the right foot into the third thoracic ganglion, viewed from below. The dashed line indicates the extent of the neuropile into which sensilla on different parts of the right leg project. Only the anterior section of the right half of the ganglion is drawn. (*a* redrawn after Siegler & Burrows, 1986; *b* from Newland, 1991; reprinted by permission of Wiley-Liss, Inc., a subsidiary of John Wiley & Sons Inc.)

which consist of a sensory cell that responds phasically when its hair (which can be up to 0.8 mm long) is deflected towards the centre of the body. Other types of sense organs, such as campaniform sensilla (see Chapter 4), monitor strains in the cuticle or span joints, responding whenever the leg is moved. Sensory neurons project in an orderly manner into the ganglion, so that their terminal branches form a somatotopic map of the surface of the leg (Newland, 1991). All the hair-like sensilla project to a particular region of neuropile in the ventral part of the third thoracic ganglion. A sensillum near the foot projects to a relatively posterior region of this neuropile (Fig. 8.5*b*), whereas a sensillum close to the joint of the femur with the base of the leg projects more anteriorly.

8.7 Local spiking interneurons

Each local spiking interneuron has two layers of branches (Fig. 8.6*a*). Studies with the electron microscope have shown that the ventral layer

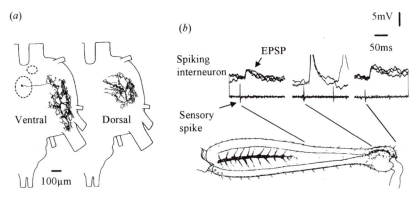

Figure 8.6 Local spiking interneurons of the locust third thoracic ganglion. (*a*) Diagrams of an interneuron showing its dorsal and ventral fields of branches. This interneuron was excited by hairs on the ventral surface of the tibia near to the foot, and by proprioceptors that responded to extension of the tibia. The two oval shapes indicate locations of cell bodies of two different groups of spiking local interneurons. (*b*) Intracellular recordings of the responses by a local spiking interneuron to stimulation of trichoid sensilla at three different locations on the femur, indicated in the diagram. Two or three recordings for each hair are superimposed. The lower trace shows a spike from the sensory neuron, and the upper trace is the intracellular recording from the interneuron. The middle recording shows the strongest connection: each sensory neuron spike evoked a spike in the interneuron, whereas spikes in the other two sensory neurons evoked subthreshold EPSPs. (*a* from Burrows, 1985; reprinted with permission of the American Physiological Society; *b* modified after Burrows, 1992b; reprinted with permission from *Trends in Neuroscience*; copyright © 1992 Elsevier Science.)

mostly receives input synapses, whereas the dorsal layer mostly makes output synapses (Watson & Burrows, 1985). The two layers are connected by a long, thin neurite that may be an axon, carrying spikes from the ventral to the dorsal layer, but its small size makes it hard to confirm this experimentally. One group of local spiking interneurons have their cell bodies clustered near the middle of the ventral surface of the ganglion, and these interneurons all make inhibitory output synapses (Siegler & Burrows, 1984). Another group have their cell bodies close to the anterior connective, and these interneurons make excitatory output synapses (Nagayama, 1989).

Each local spiking interneuron responds to stimulation of basiconic and trichoid sensilla in a well-defined receptive field on particular regions of the leg (Burrows & Siegler, 1985; Burrows & Newland, 1993). Some interneurons have fields restricted to particular parts of one segment of the leg, but others have broader receptive fields that extend in both directions from a joint. The interneuron from which the recordings in Fig. 8.6*b* were made responded to stimulation of trichoid sensilla on the ventral surface of the femur and there was a gradient in the strength of connections that declined in a direction away from the body. In the most sensitive part of the receptive field, a single spike in a sensillum will often be sufficient to elicit a spike in a local spiking interneuron. However, when two sensory neuron spikes occur in quick succession, the second may only elicit a small PSP, and this change in the effectiveness of the synapses is a mechanism that enhances responsiveness by the locust to new, rather than maintained, tactile stimuli (Burrows, 1992a).

Each spiking interneuron excites or inhibits a number of different targets, which are mainly non-spiking interneurons and motor neurons. The overall pattern of connectivity is diffuse, because each non-spiking interneuron or motor neuron receives inputs from a number of different local spiking interneurons. However, the pattern of connections preserves information about the location of tactile stimuli on the surface of the leg, so that non-spiking interneurons and motor neurons, as well as the local spiking interneurons, have well-defined receptive fields. When part of the leg is touched, therefore, a particular set of local spiking interneurons and another set of non-spiking interneurons are activated.

8.8 Non-spiking interneurons

The non-spiking interneurons (Fig. 8.7*a*) provide a major source of input to leg motor neurons. As their name suggests, these interneurons do not spike when they are excited, but changes in membrane potential directly affect the rate at which neurotransmitter is released from their output synapses. There is a smoothly graded relationship between the membrane potential of a non-spiking interneuron and of a motor neuron that it drives (Burrows & Siegler, 1978), which is seen when one electrode is used to inject current into a non-spiking interneuron while another one is used to record the membrane potential of a postsynaptic motor neuron. The same relationship between presynaptic and postsynaptic potentials is found at all chemical synapses, but the all-or-nothing nature of a spike often obscures it. In the recordings shown in Fig. 8.7*b*, recordings were made from two different motor neurons at the same time. The interneuron inhibited the flexor

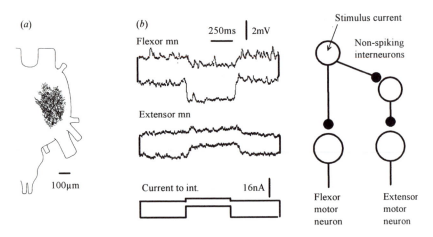

Figure 8.7 Non-spiking interneurons in the third thoracic ganglion of a locust. (*a*) The morphology of a non-spiking interneuron that excites the slow extensor tibiae motor neuron. (*b*) Intracellular recordings that show the graded nature of transmission from a non-spiking interneuron. One electrode was used to inject pulses of depolarising current into an interneuron, while a second electrode recorded the intracellular potential of a motor neuron of a leg extensor muscle, and a third electrode recorded from a motor neuron of a leg flexor muscle. When the strength of current increased, the potential changes in the two motor neurons also increased. As the circuit on the right indicates, the non-spiking interneuron probably made a direct, inhibitory connection with the flexor motor neuron, and excited the extensor motor neuron by reducing tonic inhibition from another non-spiking interneuron. (*a* from Watkins, Burrows & Siegler, 1985; reprinted by permission of Wiley-Liss, Inc., a subsidiary of John Wiley & Sons Inc.; *b* modified after Burrows, 1989.)

motor neuron and excited the extensor, probably by disinhibition of a second, inhibitory non-spiking interneuron (cf. section 7.6).

Many of the non-spiking interneurons release neurotransmitter tonically, exerting a steady synaptic drive on their postsynaptic targets. One important result is that they regulate the passage of sensory information to motor neurons, illustrated as follows for a pathway in which a sensory neuron excites a non-spiking interneuron which, in turn, excites a motor neuron. If the non-spiking interneuron is relatively hyperpolarised as a result of a particular combination of sensory inputs, its membrane potential will be below the threshold for transmitter release and its output synapses will effectively be switched off. As a result, exciting the sensory neuron to produce a spike will cause an EPSP in the interneuron, but this will be too small to cause the interneuron to release neurotransmitter and excite the motor neuron. If the angle of the joint is altered, the new combination of sensory inputs onto the interneuron might excite it sufficiently for it to release transmitter tonically, which will excite the motor neuron with a steady, depolarising potential. Now if the sensory receptor is excited, the EPSP it causes in the interneuron will be passed on to the motor neuron (Burrows, 1979). The non-spiking interneurons, therefore, can act as summing points, where signals including those from proprioceptors, tactile sense organs, and interneurons that co-ordinate activity between different segments are integrated before being passed on to the motor neurons. The result is that local reflex movements are modulated according to behavioural **context.** Interestingly, when the angle of a joint changes, the size of the change in membrane potential that occurs in non-spiking interneurons depends on whether the joint angle has increased or decreased (Siegler, 1981). This means that the properties of the interneuron and its output synapses depend upon the recent history of movements the locust has made.

8.9 Organisation of neurons that control reflex movements

A summary of the main patterns of interconnections found among neurons controlling local reflexes of a locust's leg is drawn in Fig. 8.8. Some non-spiking interneurons make excitatory output connections and others make inhibitory connections, but none has been found to excite some of its targets and inhibit others. As with the local spiking interneurons, the pattern of output connections from non-spiking interneurons is diffuse in that a non-spiking interneuron can control several different motor neurons and each motor neuron is controlled by several non-spiking interneurons (Burrows, 1980). Some non-spiking interneurons connect laterally with others, always making inhibitory connection, which might help to ensure that incompatible movements involving antagonistic muscles do not occur.

The most intriguing aspect of the organisation of the neurons that control local reflexes of a locust's leg is that one type of interneuron operates with spikes and the other without them. The situation contrasts with the way information is processed in visual systems, where the first layers of neurons often operate without spikes but later neurons transmit trains of spikes along axons. It is interesting that all the interneurons that have been found to drive flight motor neurons in locust thoracic ganglia

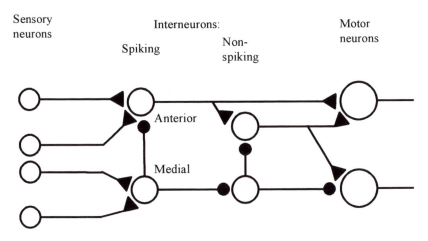

Figure 8.8 A circuit diagram indicating the kinds of connections that have been discovered between neurons involved in local reflexes of a locust's hind leg. Excitatory connections are indicated by triangles, and inhibitory connections by circles.

produce spikes (see section 7.6). It might be that the graded way in which non-spiking interneurons control motor neurons is useful for the kinds of small postural adjustments that legs can make. Spiking rather than non-spiking communication is not simply related to neuronal size, because both the extent and the complexity of the branching pattern are similar in non-spiking and spiking local interneurons (compare Fig. 8.6*a* with Fig. 8.7*a*). It is possible that signals other than spikes cannot travel along the narrow, constricted neurite that connects the dorsal and ventral fields of a spiking local interneuron. Spikes may also provide a mechanism for synchrony by depolarising all of a neuron's presynaptic terminals together and by a similar amount. In contrast, the use of smoothly graded membrane potentials in a non-spiking interneuron probably allows different regions of the neuron to operate independently to some extent. In support of this idea, electron microscopy has shown that input and output synapses are intermingled along all the processes of non-spiking interneurons and no separate regions of input or output can be distinguished (Watson & Burrows, 1988), and signals do seem to reduce in amplitude as they travel from one part of a non-spiking interneuron to another (Burrows & Siegler, 1978).

8.10 Conclusions

All three studies in this chapter have described interneurons that are not dedicated to particular behaviours, but which operate as parts of ensembles in the interface between sensory analysis and motor control. Whenever an *Aplysia* retracts its gill, many interneurons are active no matter whether the movement is a reflex in response to a sensory stimulus or a movement associated with respiratory pumping. A similar conclusion applies to the interneurons involved in local bending movements of the leech because most of them respond to touch anywhere on the surface of their segment. In the locust, both local spiking and non-spiking interneurons have distinct receptive fields in their sensory responses, so are probably more restricted in their patterns of activity. Experiments in the leech and locust involving paired recordings with microelectrodes in order to construct circuit diagrams of the neurons involved have shown that information mostly flows through one or two layers of interneurons rather than directly from sensory to motor neurons. Each interneuron sums inputs from a number of presynaptic sources and distributes outputs onwards to many different targets. Information flows in one direction, with no feedback from the motor neurons to the interneurons. Although the pattern of connections made by interneurons is complex, the movements controlled by each motor neuron are oriented appropriately in response to stimulation of an area of skin.

It is difficult to appreciate how networks of neurons organise well-coordinated behaviour simply be examining a diagram of the connection pattern. Computer modelling can be very useful because factors such as the strengths of different connections can be incorporated, and because an experimenter can trace the way that information flows sequentially through the network. Also, as the study with the leech showed, modelling with computers can test whether a particular network works in a specific way.

There is undoubtedly a continuum in the degree to which interneurons are dedicated to particular behaviours, with some giant interneurons (see Chapter 3) being apparently concerned with just one behaviour. Probably the major advantage in using neurons that are not strongly dedicated to particular behaviours is that the circuits are flexible, so that different activities in which an animal is engaged are properly integrated together. Although we can neatly categorise different activities performed by an animal as different patterns of behaviour, the diffuse nature of the organ-isation of nervous systems makes it difficult to recognise discrete circuits that are concerned with particular behaviours.

Further reading

Posner, M.I. & Raichle, M.E. (1997). *Images of Mind*, 2nd edn. New York: Scientific American Library. A book that describes a number of ways that have been used to monitor changes in the activity of populations of neurons in the human brain, often associated with the performance of particular tasks. A good example is the technique of positron emission tomography, which has pro-vided much valuable clinical information. This and other techniques monitor changes in the flow of blood to different brain regions, providing an indication of how active they are.

Lewis, J.E. & Kristan, W.B. Jr (1998). A neuronal network for computing population vectors in the leech. *Nature* **391,** 76–9. This describes work that extends the study of the interneurons responsible for controlling local bending movements in the leech by examining how information about the exact location of a touch to the animal's surface is coded by a population of neurons.

Jellema, T. & Heitler, W.J. (1997). Adaptive reconfiguration of a reflex circuit during different motor programmes in the locust. *J Comp Physiol A* **180,** 659–69. Before a jump by a locust, both the extensor and the flexor muscles of the hind leg tibiae contract together, building up tension for sudden release causing the jump. These muscles are activated in a different pattern during walking or scratching, and this paper reports one aspect of how the circuitry is modified, by altering the strengths of synapses from proprioceptors.

9 Nerve cells and changes in behaviour

9.1 Introduction

One of the most important, intriguing aspects of animal behaviour is that it continually changes. Some of the changes are parts of the processes of development and maturation, whereas others allow the animal to learn about alterations in its environment. Learning enables an animal to make and modify predictions based on experience, for example that a particular action will be followed by a rewarding or an aversive event. The ability to change is a fundamental property of many nerve cells and their interconnections, and much recent research has focused on events at the molecular and cellular level that could underlie learning or events during the development of a nervous system. A major challenge in neuroethology is to relate these changes in cellular properties to alterations in animal behaviour.

One type of trigger for changes in behaviour is provided by hormones, which can ensure that events are initiated at particular times. Steroid hormones act by regulating gene expression and can produce modifications in the morphology of nervous systems that are correlated with changes in behaviour (Breedlove, 1992; Weeks & Levine, 1995). Polypeptide hormones, on the other hand, often act to trigger particular behaviour patterns by exciting particular target neurons. Another type of mechanism triggers learning, in which an animal forms a new association between a sensory stimulus and a motor program. This commonly occurs when a particular sequence of stimuli is received by sensory pathways, leading to changes in neuronal excitability and the strengths of synapses. In the longer term, learning can also involve alterations in the pattern of gene expression in a neuron.

9.2 Growth and metamorphosis in insects

During the life history of many animals, the body form alters as the animal grows and matures. Each stage has a different pattern of behaviours from the other stages, which means that quite profound changes can occur in nerve cells and the circuits in which they participate. The mechanisms that underlie such changes have been extensively studied in a type of hawk moth, the tobacco hornworm *Manduca sexta*. The first stage in the life cycle after hatching from the egg is the larva, which spends most of its time eating, walking and growing. The second stage is the pupa, which is almost immobile and undergoes a major reorganisation of tissues, or metamorphosis, to produce the final, adult stage. The lifestyles of all three stages are quite different and, together with changes in body form, the sensory, muscular and nervous systems are extensively remodelled. The life history of *Manduca* is punctuated by **ecdyses**, in which the animal sheds its old skin and its whole behaviour is dedicated to ensuring an orderly transition from one developmental stage to the next.

9.3 Ecdysis in a hawk moth

Towards the end of each developmental stage, the larva grows a new, soft exoskeleton underneath the old one, and the behaviour of ecdysis loosens the new skin from the old and then allows the old skin to be shrugged off. Ecdysis lasts for about six hours, and it provides an excellent example of how particular neurohormones can orchestrate a co-ordinated behaviour at a particular time during an animal's life history. The activity of a small number of neurons commits the whole animal to ecdysis.

Manduca has five larval stages, each one almost identical in form to the last but slightly larger, until it becomes a pupa. This final larval ecdysis is the best understood. The first event is inflation of tracheal air sacs, particularly in the head, with air. Next, during pre-ecdysis, muscles contract and then relax synchronously in all segments of the abdomen once every 5–10 s (Fig. 9.1*a*; Copenhaver & Truman, 1982; Miles & Weeks, 1991). At the same time, abdominal appendages called pro-legs are retracted and re-extended. Pre-ecdysis loosens the old cuticle from the new and takes about half an hour. The rhythm for pre-ecdysis is generated by a network of interneurons in the last abdominal ganglion, and a single left–right pair of long

(*a*)

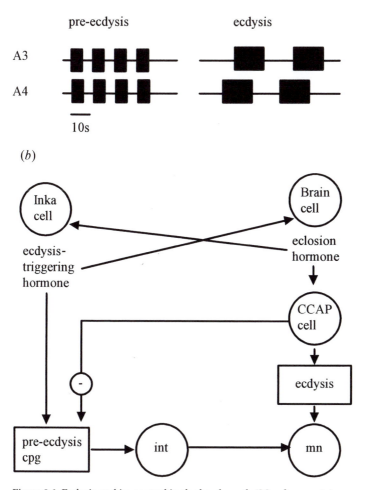

Figure 9.1 Ecdysis and its control in the hawk moth (*Manduca sexta*). (*a*) Representation of the patterns of activity that occur in pre-ecdysis and in ecdysis. Times when the same muscle in the third (A3) and fourth (A4) abdominal segments contract are shown. (*b*) A flow diagram to illustrate the actions of ecdysis-triggering hormone, released from Inka cells, and eclosion hormone, released from four brain cells. The pre-ecdysis central pattern generator (cpg) is activated by low concentrations of eclosion-triggering hormone, and the pre-ecdysis rhythm is communicated to segmental motor neurons (mn) by a pair of co-ordinating interneurons (int). Eclosion is switched on by crustacean cardioactive peptide, released from CCAP cells throughout the central nervous system, and this peptide also switches off pre-ecdysis.

interneurons communicates the rhythm to other abdominal ganglia, ensuring that contractions occur synchronously in all segments (Novicki & Weeks, 1995). Ecdysis uses the same muscle as pre-ecdysis, but in a different motor pattern in which a wave of contraction progresses anteriorly from segment to segment (Fig. 9.1a; Weeks & Truman, 1984). The neurons responsible for generating the ecdysis motor program have not yet been located.

Two polypeptide hormones are responsible for triggering and orchestrating pre-ecdysis and ecdysis (Fig. 9.1b; Zitnan et al., 1996; Gammie & Truman, 1997). First, ecdysis-triggering hormone is released into the blood from cells called Inka cells which are situated near to the openings of the respiratory tracheal system to the outside. The pattern generator for pre-ecdysis is very sensitive to ecdysis-triggering hormone and is initiated as soon as the concentration of this hormone starts to rise. Another effect of ecdysis-triggering hormone is to excite four cells in the brain that release the second polypeptide, eclosion hormone. These cells were identified with a stain for the RNA for eclosion hormone. Their axons extend right down the nerve cord to the hind end of the animal and then run anteriorly in nerves that lie alongside the gut. Activation of these cells depends on levels in the blood of the steroid hormone ecdysone, the hormone that acts as a master switch for the genes that instruct the production of new body structures during metamorphosis. Eclosion hormone is released both within the central nervous system and into the blood, and the Inka cells are excited by it, so there is a feed-forward system between the Inka cells and the eclosion hormone-releasing brain cells. Once activated, this mutual excitation is unlikely to stop, ensuring the animal is fully committed to ecdysis. The Inka cells and the brain cells discharge all of their hormone within about half an hour.

In the central nervous system, the targets for eclosion hormone are a set of 50 nerve cells, scattered fairly uniformly among the segmental ganglia. The effect of eclosion hormone is to excite these cells, by way of an intracellular messenger pathway involving cyclic GMP, and to cause them to release another peptide hormone (called crustacean cardioactive peptide). This peptide ensures the correct change in motor program by activating the pattern generator for ecdysis and at the same time switching off the pattern generator for pre-ecdysis. Ecdysis is sustained after the initial intense release of ecdysis-triggering and eclosion hormones because the increase in cyclic GMP is quite slow in the cells that release crustacean cardioactive peptide and they maintain a steady level of discharge. These cells are also excited by sensory cells that signal the continued presence of the old cuticle.

9.4 Remodelling neurons during metamorphosis

The lifestyle and behaviour of the adult moth are quite different from those of the larva, and the two stages have completely different locomotory and sensory systems. The larva moves by crawling movements involving stubby pro-legs on the abdomen and thorax, and these legs and their muscles disappear during the pupal stage. Almost all of the larval muscles die during metamorphosis, and are replaced by new muscles that move adult structures such as wings and thoracic legs. The adult has more sophisticated sense organs than the larva, such as compound eyes and antennae. Rather than being dismantled and rebuilt, the central nervous system is extensively remodelled. A few new neurons are born and some die, but many are remodelled and incorporated into new circuits.

One neuron that is remodelled is MN1, a motor neuron found in each abdominal ganglion (Levine & Truman, 1982). In the larva, the muscle it controls bends the abdomen sideways. This muscle dies during metamorphosis, and MN1 comes to innervate a new muscle that bends the abdomen upwards. This means that the left and right MN1 neurons are often used in opposition to each other in the larva, but are used together in the adult. Associated with the change in target muscle are changes in the morphology of MN1 and its connections with other neurons (Fig. 9.2). In the larva, dendrites of the neuron are nearly all on the same side of the ganglion as its axon, but it sprouts new dendrites during the pupal stage, and has a separate dendritic region on each side of the ganglion in the adult.

In the larva and in the adult, MN1 is excited by a stretch receptor which responds whenever the muscle controlled by MN1 is stretched. This stretch receptor probably makes a direct, excitatory connection with MN1 (the unbroken arrow in Fig. 9.2) because each sensory spike evokes an EPSP in MN1 after a short delay. Therefore, there is a short feedback pathway that excites MN1 whenever its muscle is stretched. The stretch receptor on the other side of the segment also participates in a circuit with MN1, but this

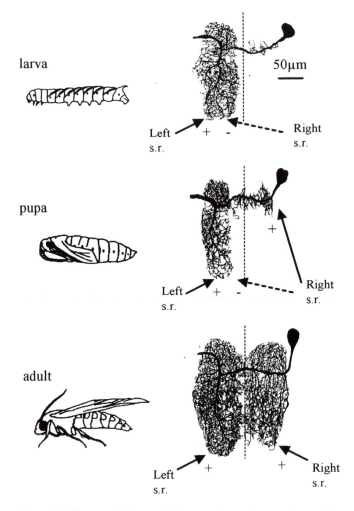

Figure 9.2 Changes in the morphology and central connections of an identified motor neuron, MN1, in *Manduca*. The MN1 drawn innervated a muscle on the left side of an abdominal segment but its cell body is on the right side of the ganglion, which is a relatively unusual arrangement in an insect. Vertical dotted lines indicate the midline of the ganglion. Arrows beneath each diagram indicate connections made by identified stretch receptor cells on the left and right sides of the animal. (Drawings of larva, pupa and adult modified after Levine & Weeks, 1990; MN1 modified after Truman & Reiss, 1988; reproduced with permission; copyright Society for Neuroscience.)

circuit alters during metamorphosis. In the larva, a spike in the right stretch receptor causes an IPSP in the left MN1. The delay between the spike and the IPSP is longer than the delay for the EPSP, so the circuit probably includes an interneuron, and is indicated by a broken arrow in Fig. 9.2. The different inputs from the two receptors help to ensure that the larva tends to straighten its abdomen if it becomes bent, and that the left and right MN1 neurons act in opposing ways.

During metamorphosis, the stretch receptor continues to make connections with the left and right MN1 motor neurons. It continues to excite the MN1 on the same side of the segment, but its inhibitory connection with the opposite MN1 becomes weaker and is then replaced with a direct excitatory connection. What is almost certainly occurring is that the new dendrites of the motor neuron come into contact with the right stretch receptor, and excitatory synapses form between the two neurons. In the adult, therefore, the left and right stretch receptors are excited whenever the abdomen is bent downwards, and act in a new proprioceptive reflex that excites the left and right MN1s. Up and down movements of the abdomen occur during flight-steering manoeuvres, during mating and during egg laying.

The growth of new dendrites of MN1 occurs at the same time as a rapid rise in the level of ecdysone within the body during the pupal stage (Truman & Reiss, 1988). Experiments on diapausing pupa, in which the emergence of the adult can be delayed for months in response to environmental stress, have shown that it is ecdysone that causes the motor neuron to grow new dendrites. In diapausing pupae, the concentration of ecdysone is unusually low, and MN1 does not grow new dendrites. Injecting ecdysone into a diapausing pupa causes MN1 to sprout its new dendrites.

The effect that ecdysone has on MN1 is regulated by another hormone, a turpenoid called juvenile hormone. If juvenile hormone is present when ecdysone levels rise, MN1 retains its larval characteristics. The same combination of rising levels of ecdysone and low levels of juvenile hormone has widespread effects on the central nervous system and elsewhere in the body (Streichert & Weeks, 1995; Levine, Morton & Restifo, 1995). It causes many of the larval muscles and some of the motor neurons to die, as well as triggering the development of new dendrites in motor neurons that survive, such as MN1. However, not all of the events involved in remodelling the nervous system are triggered by rising ecdysone levels. For example, the inhibitory connection from the right abdominal stretch receptor to the left MN1 disappears before the rise in ecdysone, and its cause is not yet known.

9.5 Associative learning and the proboscis extension reflex in honey bees

Some of the most interesting behaviour patterns amongst insects are found in the social hymenoptera, bees, wasps and ants. These animals are capable of memorising landmarks, which they use for navigation (Dyer, 1996; Judd & Collett, 1998), and they have elaborate systems for communication amongst individuals, such as the waggle dance of honey bees (Lindauer, 1967). The disposition of bees to learn to associate colours and shapes of flowers with a good source of food was recognised by one of the earliest experimental ethologists, Karl von Frisch, and he exploited this to investigate several features of the sensory capabilities of bees. Bees remain a favoured subject for behavioural experiments because they are quite easy to obtain and to train in laboratory conditions. It is also possible to make electrophysiological recordings from single neurons in a bee's brain during the performance of some simple behaviour patterns in order to reveal circuits that are important in learned behaviours.

One of the simplest behaviours that a honey bee (*Apis mellifera*) produces is to extend its proboscis, or tongue, in response to a drop of sucrose solution applied to chemosensory hairs on the proboscis or antenna (Fig. 9.3a). A bee will occasionally extend its proboscis without any obvious stimulation, or if a tiny puff of a particular odour such as the smell of carnation or orange is blown at an antenna. If the bee has recently tasted sucrose, the likelihood that it will extend its proboscis in response to a subsequent odour puff increases – stimulation with sucrose is said to **sensitise** the proboscis-extension response. A much greater enhancement of the response to a puff of odour occurs, however, after the odour puff has been paired with delivery of a drop of sucrose. For maximum effect, the odour puff must be delivered between 1 and 3 s before the drop of sucrose (Bitterman *et al.*, 1983). The next time that the odour is directed at an antenna, there is a very high chance that the bee will extend its proboscis (Fig. 9.3b). In the bee's brain, an **association** has been made between the odour and the likely presence of sucrose. The order in which the two

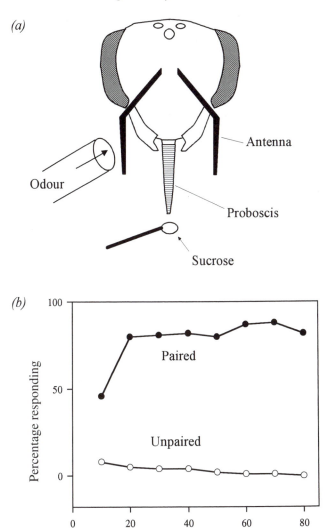

(a)

Antenna

Odour

Proboscis

Sucrose

(b)

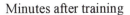

Minutes after training

Figure 9.3 Conditioning of the proboscis extension reflex in honey bees (*Apis*) to odours. (*a*) A diagram to illustrate the conditioning procedure. A puff of air laden with a chosen odour is blown across an antenna, and sucrose delivered to the proboscis when it is extended acts as the reward. (*b*) A graph to show the acquisition of the conditioned response. Groups of bees were trained by pairing a puff of odour with a sucrose reward 2 s later. After this pairing procedure, about 80 per cent of bees tested responded to the odour with proboscis extension. In comparison, few untrained bees responded to the odour. (*b* redrawn after Menzel, 1990.)

different stimuli are delivered is vital for the formation of memory. If the sucrose is applied to the proboscis just before the odour is blown over an antenna, a second puff of the odour delivered a minute later is unlikely to cause proboscis extension.

The proboscis-extension reflex is a good example of **conditioning**, a type of behavioural change often studied in vertebrates. The odour stimulus becomes conditioned so that after it has been paired with a sucrose reward, it reliably evokes a stimulus that was previously only rarely linked with this stimulus. A single pairing of a puff of carnation with a taste of sucrose is sufficient to enhance the new coupling between carnation and proboscis extension for many hours. However, the coupling will weaken and become extinguished if several odour puffs are delivered with no sucrose as a reward, and can be replaced by a new association between another odour, such as orange, and proboscis extension. The memory of the association between an odour and sucrose goes through different phases in time, so that the initial short-term memory is transferred into longer-lasting medium-term and long-term memories (Hammer & Menzel, 1995).

9.6 Neuronal pathways and conditioning

Taste-sensitive sensory neurons that detect sucrose project from the proboscis into the suboesophageal ganglion, which is the first ganglion in the nerve cord after the brain and is also the ganglion that contains the motor neurons of the proboscis muscles. Odours are detected by sensilla on the antennae, which project into antennal lobes of the brain. After processing in the brain, information about odours is carried to the suboesophageal ganglion by neurons that originate in the forebrain, or protocerebrum. During conditioning, information about the sucrose reward must be carried to the brain from the suboesophageal ganglion and cause modification of the pathways that link olfactory sensory receptors with the protocerebral neurons. The pathways involved are shown in Fig. 9.4.

About 60 000 olfactory receptors run from each antenna to a discrete area on either side of the brain called the olfactory lobe. Each sensory neuron terminates in a spherical structure called a **glomerulus**, and olfactory information is processed both within and between the 160 glomeruli in an olfactory lobe before being distributed by projection neurons to other brain regions. Some neurons project directly from glomeruli to the protocerebrum, and

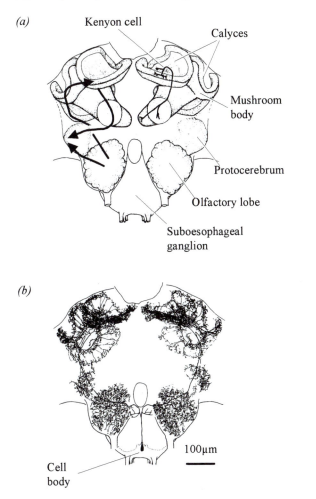

Figure 9.4 The morphology of the brain and suboesophageal ganglion of the bee to show the structures and pathways involved in conditioning the proboscis-extension reflex to odours. (*a*) Information about odours is processed in the olfactory lobes, which send outputs both to the protocerebrum and to the calyces of the mushroom bodies, where Kenyon cells have their dendrites. Kenyon cell axons have branches to two separate output lobes of a mushroom body, from where interneurons travel to various brain regions, including the protocerebrum. Some interneurons travel from the protocerebrum to the suboesophageal and the segmental ganglia. Motor neurons that control proboscis muscles are situated in the suboesophageal ganglion, which is also where sucrose-detecting chemosensory neurons terminate. (*b*) A diagram to show the extent of the innervation pattern of neuron VUMmx1, whose cell body and dendrites are in the suboesophageal ganglion. (*a* and *b* after Hammer, 1993; reprinted with permission from *Nature*; copyright © 1993 Macmillan Magazines Ltd.)

others project to another brain region, called the **mushroom body**. The mushroom bodies were named after their resemblance to some kinds of horn-shaped mushrooms, and have been implicated in the more complex types of insect behaviour, including learning and social behaviour. They are particularly well developed in bees, and also in some other types of insects, including cockroaches, where they may play roles in spatial memory (Mizunami, Weibrecht & Strausfeld, 1993). Each mushroom body contains a parallel array of neurons called Kenyon cells (one is drawn in Fig. 9.4a). About a third of all the neurons in the brain of a bee are Kenyon cells, with 170 000 in each mushroom body. It is extremely difficult to make recordings from them because they are small and tightly packed together. Their dendrites are arranged in structures called calyces, and olfactory information arrives at the outermost rim region of each calyx. Kenyon cell axons branch into two, one branch going into the α lobe of the mushroom body and the other branch into the β lobe. Various output neurons combine information from many Kenyon cells and deliver information from the mushroom bodies to other parts of the brain. Response properties of these output cells can alter during conditioning (Mauelshagen, 1993).

Two types of experiment suggest that both the antennal lobe and the mushroom body play roles in associative conditioning. In one, a small probe was used to cool local regions of the brain to 1–5 °C for 10 s, transiently blocking or reducing activity (Erber, Masuhr & Menzel, 1980). Cooling either an olfactory lobe or a mushroom body reduced conditioning, and the effect was not simply to reduce sensory activity because conditioning was blocked if several seconds or even minutes elapsed between delivery of the odour stimulus and the start of cooling. In the second type of experiment, a small amount of the neuromodulator octopamine was injected into a brain region just after delivery of an odour pulse. The octopamine substituted for the effects of stimulation of taste receptors with sucrose, and when the odour was presented later on its own, it evoked proboscis extension. Octopamine had this effect when injected into the olfactory lobe or calyx of a mushroom body, but not elsewhere.

9.7 The role of an identified neuron in conditioning

A bee's brain must be able to form associations between a sucrose reward and many different other stimuli, including a variety of odours. We could, therefore, expect that a neuron that plays a key role in conditioning would have branches distributed in many different regions of the brain. One such neuron is VUMmx1 (Fig. 9.4b), an unpaired neuron which has branches that are arranged symmetrically on both the left and right sides of the suboesophageal ganglion and brain, and Martin Hammer (1993) demonstrated experimentally that this neuron can induce conditioning. Its branches overlap the pathways that link olfactory stimuli to the extension reflex at several places. Like many other unpaired, median neurons in insects, such as DUM neurons in the thoracic ganglia (see Box 7.2, p. 187), VUMmx1 appears to contain and release octopamine.

Sucrose excites VUMmx1, and this neuron continues to spike for at least half a minute after the sucrose has been removed. VUMmx1 is also excited, but only for a brief time, by various odours directed at the antennae. However, Hammer found that when sucrose was delivered just after a particular odour, the next time the odour was delivered alone VUMmx1 responded with a brief, intense burst of spikes followed by excitation that continued for 20 s, a similar response to that given by VUMmx1 to sucrose. Like proboscis extension, this response enhancement depended strictly on the two stimuli being delivered within a short time, with the start of the odour preceding the start of sucrose delivery. The neuronal mechanism that requires a particular stimulus order is unknown.

Exciting VUMmx1 by injecting current into it through a microelectrode can substitute for stimulation with sucrose in conditioning (Fig. 9.5). In forward pairing, the odour stimulus started 2 s before the electrical excitation of VUMmx1, and during backward pairing, VUMmx1 excitation preceded the odour by 5 s. Activation of the proboscis-extension muscle was monitored in an electromyogram recording. After forward pairing, a test pulse of odour 10 min later produced a large electromyogram response (Fig. 9.5b). In contrast, backward pairing of VUMmx1 excitation with odour did not lead to any enhancement of the response to the test odour. This experiment clearly shows that VUMmx1 can condition the proboscis-extension response to a particular odour. However, it does not show that VUMmx1 is solely responsible for conditioning, and it probably normally works in parallel with other similar ventral, unpaired median neurons in the suboesophageal ganglion.

VUMmx1 can deliver information about a sucrose reward to the olfactory lobes and the mushroom body, the sites that the experiments with local

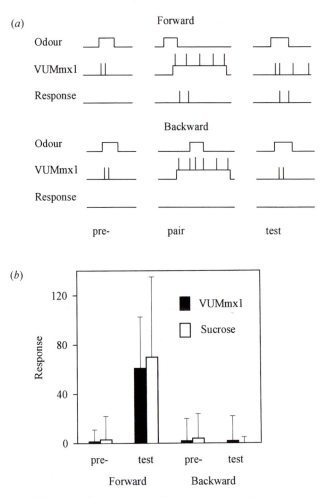

Figure 9.5 Electrical excitation can substitute for stimulation with sucrose in conditioning the proboscis-extension response to odours. (*a*) An outline of the protocol used in training experiments. First, any response by proboscis-extension muscles to a puff of odour was measured in a pre-test, Next, a puff of the odour was paired with electrical excitation of VUMmx1 by a pulse of depolarising current. In forward pairing, the odour preceded the start of the current pulse, and in backward pairing the current pulse started before the odour. Finally, responses to a test pulse of odour were recorded. (*b*) Responses to a test puff of odour after training with either electrical excitation of VUMmx1 or a sucrose reward. The response was measured as the number of spikes in a proboscis extensor motor neuron during 10 s after a test or pre-test puff of odour, and the histograms plot the median response from groups of 11–15 bees. The error bars indicate the maximum responses from each group. (*a* and *b* redrawn after Hammer, 1993.)

cooling and octopamine injection showed are important in conditioning. A key event in conditioning is that excitation of VUMmx1 is closely associated in time with stimulation of particular odour-activated pathways. A great deal of information about the molecular mechanisms for coincidence detection in neurons has been gained in studies on the gill-withdrawal reflex of the mollusc *Aplysia* (Box 9.1). The mechanism by which VUMmx1 establishes conditioning is likely to involve intracellular messenger molecules such as cyclic AMP and cyclic GMP and, in *Drosophila*, mutations that affect production of these molecules also cause impairment of memory (Belvin & Yin, 1997). Conditioning of proboscis extension in bees is quite complex because a large range of possible stimuli can become associated with the response, and it is probable that the complexity of the mushroom bodies is in some way related with the need to be able to deal with a large number of possible stimulus configurations.

9.8 Bird song and its production

Almost all birds produce calls of various kinds but one group of passerines, 4000 species called song birds, produce extended and complex songs. A defining characteristic of these songs is that their full expression requires learning by a juvenile bird from mature, singing adults that act as tutors. Song birds are believed to have evolved from one common ancestor, and include finches, warblers and thrushes. The most complex songs are usually produced by males during the breeding season, and the functions of song include establishing territories, attracting potential mates, and maintaining pair bonds (Catchpole & Slater, 1995). Because birds incorporate parts of the songs of other birds into their own songs, local dialects develop in some species (Marler & Tamura, 1964; Baker & Cunningham, 1985).

For a neuroethologist, bird song provides some general lessons about the way a complex type of behaviour is controlled by a central nervous system. Within the brain, a number of discrete areas, or **nuclei**, have been shown to be involved in song. Each nucleus contains cell bodies and dendrites of a number of types of neuron. Some of the neurons have axons that run in tracts to other nuclei, and other neurons participate in processing information within their nucleus. One group of nuclei is responsible for generating the song in the adult, and another is involved in laying down motor

Box 9.1. Plasticity in the gill-withdrawal reflex of *Aplysia*

Aplysia withdraws its delicate gill if skin near to the siphon is touched (see section 8.2). Properties of the synapse that links a sensory and motor neuron in the abdominal ganglion (∗) change as a result of different kinds of sensory experience, and contribute to simple plastic changes in behaviour. If the synapse is activated repeatedly, the strength of transmission declines and this is one mechanism underlying **habituation** of the gill-withdrawal reflex. A noxious stimulus applied elsewhere to the body, such as to the tail, sensitises the reflex so that the next siphon stimulus will evoke a larger gill-withdrawal response than previously. Sensitisation is due to the action of facilitating interneurons which release the transmitter serotonin onto the presynaptic terminals of the siphon sensory neurons. This causes **presynaptic facilitation**, an increase in the amount of neurotransmitter released (Byrne & Kandel, 1996). The circuit can also be conditioned by following a touch to the siphon immediately by a noxious stimulus to the tail. After this pairing, the next touch to the siphon causes a greatly enhanced postsynaptic potential in the gill-withdrawal motor neuron. At least two coincidence-detecting mechanisms operate to enhance synaptic activity. The first is activated when serotonin stimulates the sensory terminal of the sensory neuron while the sensory neuron is electrically excited, and this increases the manufacture of cyclic AMP in the sensory terminal (Hawkins, Kandel & Siegelbaum, 1993). The second is similar to a phenomenon known as **post-tetanic potentiation**, familiar in some parts of the vertebrate central nervous system. It occurs if the sensory neuron is stimulated at a time when the motor neuron is already excited, and involves a postsynaptic receptor protein called an NMDA receptor (Glanzman, 1995).

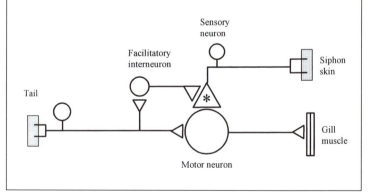

programs for singing. The subject of many studies is the zebra finch (*Taeniopygia guttata*), an Australian species that breeds readily in captivity and develops to maturity in only 100 days. An individual male produces song that is more easy to characterise than the songs of many other birds because it is relatively stereotyped.

A lot of information about a song can be expressed in a sonograph, in which records of sounds are broken down to show the relative contributions of different frequencies. Sonographs are particularly useful for comparing the songs of different individuals, or of one bird at different stages in its development. The sonograph in Fig. 9.6*a* illustrates different levels of organisation within the song of a zebra finch. Series of notes are linked together into discrete syllables, and a series of syllables is linked together in a unit called a motif. When birds are interrupted during singing, they always finish a syllable (Cynx, 1990), which means that a syllable is a basic unit in the organisation of song. In zebra finches, the motif that a particular male sings is fixed in form, although the number of motifs in a bout of singing varies from song to song, as do the brief introductory notes that precede the song and separate successive motifs. Each note of the zebra finch song is composed of sounds of many different frequencies. Notes of most other species contain more restricted tones and, as a result, have a more musical quality to a human listener.

The organ responsible for producing sounds during song is the **syrinx**, located where the trachea joins the bronchi of the two lungs (Fig. 9.6*b*). Four to six muscles on either side are attached to the syrinx, and sound is produced when air is expelled through it. The exact way in which sound is produced has not been conclusively demonstrated, but it may involve the openings of the bronchi into the trachea forming whistles, or the two medial tympaniform membranes vibrating like drum membranes. Some syrinx muscles determine the tone of sound which is produced and others control the timing of sounds by opening and closing the bronchi. Respiratory muscles generate the force for expelling air through the syrinx, and so control the volume of sound. The lungs of birds are rigid and air moves through them in one direction by the action of large air sacs that act as bellows. Each syllable of a song is produced by contraction of muscles that expel air from the interclavicular air sac. Electromyograms and pressure measurements have shown that, in canaries, each syllable is co-ordinated with a cycle of inspiration and expiration, even at rates in excess of 20/s. It is

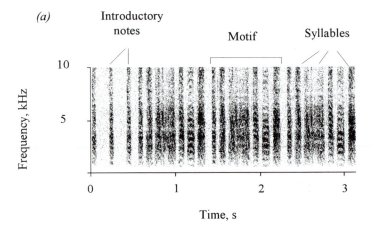

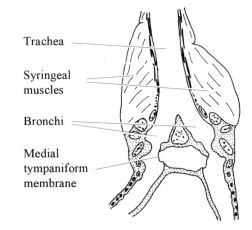

Figure 9.6 Bird song and its production. (*a*) A sonograph of the song of a zebra finch (*Taeniopygia guttata*). Shown here are a few introductory notes, followed by three repetitions of the same motif. This motif contained six syllables, each one of which was a particular sequence of notes. (*b*) The main structures associated with the syrinx of a song bird. (*a* sonograph kindly supplied by Dr D. Margoliash; *b* from Suthers, 1990; reprinted with permission from *Nature*; copyright © 1990 Macmillan Magazines Ltd.)

usual for the muscles on the left and right of the syrinx to act independently of each other, and zebra finches sing mostly by using muscles on the right.

9.9 The development of song

The development of song in most species follows three distinct phases. In the first, the sensory phase, the bird hears songs produced by tutor birds around him. In temperate species, this usually occurs soon after hatching in the spring or summer, almost a year before the bird begins to sing himself. The requirement for tutor songs was first shown experimentally by Thorpe (1958), who found that chaffinches (*Fringilla coelebs*) that had been reared in the laboratory without hearing the songs of adults produced very abnormal songs when they matured. The syllables of their songs were simple and disorganised. However, chaffinches that had heard tape recordings of adult chaffinch song as fledglings produced normal songs when they matured. Young chaffinches did not learn from other sounds, including the songs of most other bird species, so a mechanism for recognising the song of members of its own species must be built into the brain of the young bird. The timing of exposure to tutor song is critical. A young chaffinch must hear adult songs before the middle of its first summer if it is to sing normally the next year.

The second period of song development, the sensorimotor phase, begins early in the spring of a chaffinch's second year. The bird starts to sing spontaneously, at first producing quiet and variable songs. Gradually, song becomes louder and, because it is variable, it is called plastic. Plastic song is often a mixture of two different types of song, and includes repeated syllables. A bird must be able to hear its own plastic song for normal development (Konishi, 1965a, 1965b). In the final period, in late spring, song crystallises to its mature form, probably triggered by a rise in the level of the steroid hormone testosterone in the blood. In most species, deafening at this stage does not cause abnormalities in song.

This sequence of three phases is found in most species of song bird, but there are many variations in detail. Some species, including starlings and canaries, are called open-ended learners because they go through the processes of song practising and crystallisation every season. Zebra finches follow the general pattern, except that young males can still hear the songs of tutors while they start to produce plastic song, which is associated with their rapid development and social lifestyle. Their sensitive period for song learning starts at 20 days old, when they fledge, and lasts until they are 40 days old. Young zebra finches begin to sing 25 days after hatching and song crystallises 90–110 days after hatching, at sexual maturity.

9.10 Neural centres for hearing and singing

Three different groups of nuclei are known to play roles in the production of song. One group of brain nuclei is involved in processing auditory information, a second in producing the motor program for song (Fig. 9.7a), and a third in learning and developing the song pattern (Fig. 9.7b). One nucleus, HVc, is a member of all three groups. Sometimes, HVc is referred to as the 'higher vocal centre', but attaching functional names to brain regions can be misleading, and this and other nuclei are referred to here by accepted abbreviations. In an adult male zebra finch, HVc contains about 35000 neurons. There is a good correlation between the size of HVc and the complexity of song in different species.

HVc exerts direct control over the production of song through connections it makes with nucleus RA. From RA, axons project to groups of motor neurons that innervate the syrinx and some respiratory muscles. Birds in which HVc or RA has been destroyed cannot sing, although they still court females by adopting the same postures as during singing (Nottebohm *et al.*, 1976), and they produce alarm cries and some other calls. Anthony Yu and Daniel Margoliash (1996) implanted fine wire electrodes into the brains of zebra finches and recorded spikes from single neurons while the birds were moving around their cages and singing normally. Many of the neurons in HVc started to spike before a bird began to sing, and remained excited until just before the end of the song. During a song, spike rate increased and decreased in a characteristic manner whenever the bird sang a particular syllable. A particular neuron was most excited when a particular syllable was sung, and each syllable would be associated with excitation of a unique population of HVc neurons. In contrast, individual neurons in RA produced discrete bursts of spikes coinciding with a specific series of notes that occurred in a number of different syllables throughout the song. In other experiments, Eric Vu and colleagues (Vu, Mazurek & Kuo, 1994) used similar electrodes to stimulate neurons in singing birds with a short train of pulses that lasted less than the duration of a syllable. Stimuli to small regions of

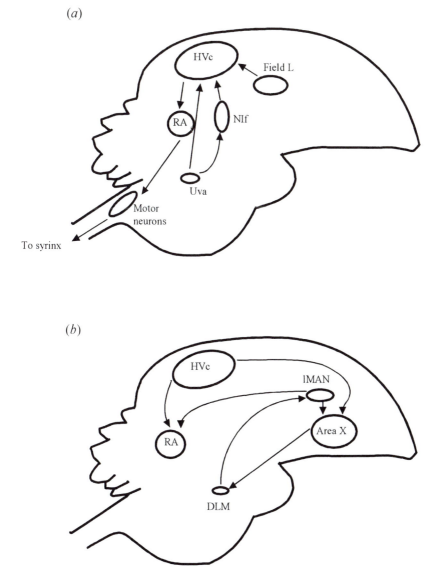

Figure 9.7 Diagrams to show the principal brain nuclei involved in the generation and development of bird song. (*a*) The control of singing in a mature adult. Motor neurons of syringeal muscles are activated by neurons from RA which, in turn, is activated by neurons from HVc. HVc receives inputs from Field L, which is a major sensory area involved in auditory processing, and from Uva and NIf. (*b*) Nuclei involved in the development of song. Notice that nucleus IMAN receives feedback from HVc via Area X

HVc disrupted the syllable the bird was singing and interfered with the order of subsequent syllables within a motif. For example, if a bird's motif consisted of syllables A B C D, a stimulus delivered just after syllable B might alter the motif's structure to A B G, combining syllables C and D to an unusual, abbreviated form, G. The electrical stimuli neither stopped the song nor interfered with the pattern of syllables during following motifs. Stimuli delivered to RA had a more restricted effect, disrupting the current syllable without affecting later syllables within the motif. A stimulus delivered to RA just after B in the sequence A B C D would cause the bird to sing A B C' D, where C' is an altered form of syllable C.

These two kinds of experiment suggest that individual neurons in HVc control particular sequences of syllables within a motif by recruiting groups of neurons in RA in a specific order. The RA neurons each excite a pool of motor neurons that control muscles of the syrinx to produce a particular combination of notes, and this correlates well with the anatomy of RA, which is arranged topographically so that a particular region in it corresponds with a particular group of motor neurons and muscles (Vicario, 1991). HVc is not organised topographically.

HVc has sensory as well as motor properties. In an anaesthetised bird, many HVc neurons respond well to recordings of the bird's own song, but not to the songs of other birds. The neurons recognise both the order of notes within a syllable and the order of syllables within a motif. Specificity for the order of syllables in a motif is particularly interesting because it means the HVc neurons need to collect information over time, for example to distinguish order A B C D from A C B D. It is not clear whether these auditory responses play a role in the control of singing in the adult because they are probably suppressed when the bird sings (McCasland & Konishi, 1981), but they are important for the normal development of song.

9.11 Development of song nuclei

Nuclei lMAN, DLM and Area X are required for a bird to acquire the normal adult song pattern (Bottjer, Meisner & Arnold, 1984), but not for an adult bird to sing. If lMAN is removed from a fledgling bird, the bird will produce an incomplete pattern of song, and if Area X is removed, song remains plastic. In the adult, neurons in Area X and lMAN, like those in HVc, respond specifically to tape recordings of the bird's own song. This selectivity arises gradually. Neurons in lMAN and Area X of young birds are excited by many kinds of sounds, including songs played in reverse, and do not prefer particular songs (Doupe, 1997; Solis & Doupe, 1997). In 60 day-old zebra finches, some of the neurons prefer the bird's own song, whereas others respond best to tutor songs. When song crystallises, all the neurons prefer the bird's own song. The gradual nature with which these auditory responses develop supports a mechanism of instruction during song development. The less likely alternative is that during the sensorimotor phase, particular song-producing circuits are selected from an array of alternative circuits that are already fully formed in the young bird. If development occurred by selection, it would probably be abrupt rather than gradual.

The anatomical organisation of the song nuclei changes considerably during development. HVc is first recognisable as a distinct nucleus 10–15 days after hatching, and by day 50, half way through sensorimotor learning, the number of neurons it contains increases by 50 per cent (Kirn & DeVoogd, 1989). The neurons that are added to HVc are new cells that originate from a proliferative layer in the ventricle above HVc. In another bird, the canary, new neurons are added to HVc every year (Box 9.2). During sensorimotor learning, RA also increases in size, but lMAN loses about half its neurons. To begin with, 37 per cent of the synapses in RA are from lMAN neurons, but this drops to 4 per cent by the time song crystallises (Herrmann & Arnold, 1991).

Box 9.2. Seasonal birth of new neurons in the canary brain

In all animals, most nerve cells are present at birth, and it is well known that in adult humans brain cells that are lost through injury are not replaced. It was surprising, therefore, when Goldman and Nottebohm (Nottebohm, 1989) demonstrated that, in canaries, new neurons are added to HVc. These neurons are born in the brain cavity, or ventricle, and migrate along glial cells to reach HVc. Some of these new neurons grow axons that extend several millimetres to RA. The rate at which new neurons are added to HVc is highest in the autumn, which is when adult male canaries learn new songs each year. Although there seems to be an association between the development of new songs and the annual birth of new neurons, female canaries also add some new neurons to their HVc, as do adult zebra finches, and it has now been shown that a few new neurons are born in the brains of adult mammals.

The essence of the development of song is that a motor program is modified by comparing the sound of the bird's own song with the memory of aspects of tutors' songs. One potential site for comparisons is Area X, where some signals arrive directly from HVc and others are fed through a loop that takes them from Area X and returns them there after travelling through DLM and lMAN (Fig. 9.7*b*). Another potential site for comparisons is RA, which receives information from HVc both directly and through lMAN (Fig. 9.7*b*). Clearly, lMAN plays a key role in the developing song-control circuitry. Individual neurons from lMAN send axonal branches to both Area X and RA (Vates & Nottebohm, 1995), and lMAN is the nucleus that shows the greatest loss of neurons during song development. Perhaps as the bird improves his song, particular neurons in lMAN may be selected to survive and maintain their contacts, whereas other lMAN neurons die.

The feedback loops evident in Fig. 9.7*b* may function to delay copies of the motor commands issued by HVc, so that they coincide in time with sensory responses to the sound of the bird's song (Margoliash, 1997). A motor command issued by HVc precedes the sound of a syllable by at least 50 ms; and the response to the sound of the syllable does not arrive in HVc until at least 20 ms after the sound. The brain could compare the motor and sensory signals if the two signals were brought together at particular nuclei, with a longer route for the copy of the motor command issued by HVc and a shorter route for the auditory signal. In response to particular syllables of its own song, many HVc neurons in an individual bird spike intensely and synchronously, which would be a powerful trigger for change. When two signals coincide in time, they might trigger changes in the excitability of neurons, or in the strengths of particular synapses, as occurs during other forms of learning. It has been found that the act of singing causes the activation of a particular set of genes in some brain nuclei (Jarvis & Nottebohm, 1997), so it is likely that morphological changes, perhaps involving synapses, follow particular patterns of electrical activity.

9.12 Conclusions

Nervous systems are programmed to ensure that behaviour changes in response to particular events during the normal life of an animal. Some changes are part of the developmental program of an animal, and others allow it to adapt its behaviour when its external environment alters. A good example of how a series of events is orchestrated at a particular time during development is provided by ecdysis in moths. The larva becomes fully committed to ecdysis by the positive feedback loop in which two different polypeptide hormones, ecdysis-triggering hormone and eclosion hormone, reinforce the release of each other. Pre-ecdysis is the first motor activity to be triggered because its pattern generator is very sensitive to ecdysis-triggering hormone and is switched on as soon as this hormone starts to be released. Eclosion hormone switches off the pre-ecdysis behaviour and, at the same time, switches on ecdysis. It mediates its effects by exciting cells that release crustacean cardioactive peptide, and these remain excited for some time after their initial stimulation by eclosion hormone, until ecdysis is complete.

Polypeptide hormones exert their effects on neurons by binding with protein receptors on the cell surface and triggering intracellular messenger pathways. Steroid hormones, on the other hand, operate directly in the cell nucleus, regulating gene expression. Before each moult in a moth, the level of the steroid hormone ecdysone rises. If this rise occurs when the titre of juvenile hormone is low, it can trigger widespread effects that play an important role in the reorganisation of the nervous system, sense organs and muscles that occurs during metamorphosis. The rise in ecdysone causes some neurons and muscles to die, and others, such as MN1, to grow new dendrites which enable them to participate in new circuits with other neurons.

Less persistent changes occur during associative learning, for example when a honey bee is conditioned to associate a particular odour with a sucrose reward. The trigger for the reinforcement of specific neuronal circuits is a particular pattern of activity, which happens when sensory signals about a specific stimulus, such as an odour, just precede signals about the sucrose reward. An identified neuron, VUMmx1, has been shown to carry information about the reward and to be able to reinforce the association between a specific odour and proboscis extension. This neuron branches to many brain regions and could reinforce a large number of different sensory stimuli, which is important so that the bee can make use of many types of sensory cues to increase its foraging efficiency. During conditioning, two areas where the action of VUMmx1 is vital are the antennal lobes and the mushroom bodies. The large numbers of neurons in the mushroom bodies probably enable the bee to make novel association between many different sensory stimuli and particular actions. The wide range of possible associations that animals can make during learning provides a major challenge to neurobiologists.

The process of development of a male song bird's brain includes a program which ensures that it stores memories of the songs of adult tutors and uses the memories to mould its own, individual song pattern. There is no obvious immediate pressure for a young bird to learn songs, but learning is vital because young males that are prevented from using the songs of mature, singing adults as models for their own songs are unable to attract mates. Birds do not copy the songs of tutors exactly, and an intriguing problem is provided by the origin of the characters that distinguish the song of one individual from another.

A specific set of brain nuclei is involved in the development of song, although these nuclei play no known role in the control of mature song in the adult. During development, new neurons are added to some of these nuclei, and many neurons die in others. A number of loops and feedback pathways involving the nuclei probably enable comparisons between the sound of the bird's own song and the remembered songs of tutors. Increasing proficiency at singing is accompanied by an increase in the auditory responses of neurons in song nuclei until, in the adult, the neurons prefer recordings of the bird's own song to any other sound. The gradual change in the specificity of the auditory responses supports the idea that the development of the program for song occurs by instruction rather than by selection. Despite its complexity, bird song has become one of the most active areas of research in neuroethology. This is partly because it is intrinsically interesting, but also because the brains of animals that specialise in particular behaviours, such as bird song, almost invariably prove to be good sources of information about how the nerve cells dedicated to that behaviour work.

Further reading

Dudai, Y. (1989). *The Neurobiology of Memory. Concepts, Findings, Trends.* Oxford: Oxford University Press. The book outlines clearly, and often in a thought-provoking way, many different types of learning in a variety of animals.

Ewer, J. & Truman, J.W. (1996). Increases in cyclic 3'5'-guanosine monophosphate (cGMP) occur at ecdysis in an evolutionarily conserved crustacean cardioactive peptide-immunoreactive insect neuronal network. *J Comp Neurol* **370**, 330–41.

This paper is about plasticity at an evolutionary level; it describes differences among orders of insects in some of the neurons that control ecdysis.

Heisenberg, M. (1995). Pattern recognition in insects. *Curr Opin Neurobiol* **5,** 475–81. A review of how insects learn visual patterns.

Jacobs, L.F. (1996). The economy of winter – phenotypic plasticity in behavior and brain structure. *Biol Bull* **191,** 91–100. This review outlines changes that occur in the hippocampus of the brains of birds and mammals when food resources become scarce in winter. The hippocampus shrinks every year in mammals that hibernate, but grows in birds that store food.

Wehner, R. Lehrer, M. & Harvey, W.R. eds. (1996). Navigation. *J Exp Biol* **199**. This volume includes reviews about how insects and other animals use memorised features, such as landmarks, to find routes.

REFERENCES

Aidley, D.J. (1998). *The Physiology of Excitable Cells*, 4th edn. Cambridge: Cambridge University Press.

Baader, A. (1991). Auditory interneurons in locusts produce directional head and abdomen movements. *J Comp Physiol A* **169**, 87–100.

Bacon, J. & Möhl, B. (1983). The tritocerebral commissure giant (TCG) wind-sensitive interneurone in the locust. 1. Its activity in straight flight. *J Comp Physiol* **150**, 439–52.

Baerends, G.P. (1985). Do dummy experiments with sticklebacks support the IRM concept? *Behaviour* **93**, 258–77.

Baerends, G.P. & Drent, R.H. (1982). The herring gull and its egg. Part II. The responsiveness to egg-features. *Behaviour* **82**, 1–416.

Baker, M.C. & Cunningham, M.A. (1985). The biology of bird-song dialects. *Behav Brain Sci* **8**, 85–133.

Bartelmez, G.W. (1915). Mauthner's cell and the nucleus motorius tegmenti. *J Comp Neurol* **25**, 87–128.

Basarsky, T.A. & French, A.S. (1991). Intracellular measurements from a rapidly adapting sensory neuron. *J Neurophysiol* **65**, 49–56.

Belvin, M.P. & Yin, J.C.P. (1997). *Drosophila* learning and memory: recent progress and new approaches. *BioEssays* **19**, 1083–9.

Bicker, G. & Pearson, K.G. (1983). Initiation of flight by stimulation of a single identified wind sensitive neurone (TCG) in the locust. *J Exp Biol* **104**, 289–94.

Bitterman, M.E., Menzel, R., Fietz, A. & Schäfer, S. (1983). Classical conditioning of proboscis extension in honeybees (*Apis mellifera*). *J Comp Psychol* **97**, 107–19.

Bodenhamer, R., Pollak, G.D. & Marsh, D.S. (1979). Coding of fine frequency information by echoranging neurons in the inferior colliculus of the Mexican free-tailed bat. *Brain Res* **171**, 530–5.

Bottjer, S.W., Meisner, A. & Arnold, A. (1984). Fore-brain lesions disrupt development but not maintenance of song in passerine birds. *Science* **224**, 901–3.

Breedlove, S.M. (1992). Sexual dimorphism in the vertebrate nervous system. *J Neurosci* **12**, 4133–42.

Bruns, V. & Schmieszek, E. (1980). Cochlear innervation in the greater horseshoe bat: demonstration of an acoustic fovea. *Hearing Res* **3**, 27–43.

Burrows, M. (1975). Monosynaptic connexions between wing stretch receptors and flight motorneurones of the locust. *J Exp Biol* **62**, 189–219.

Burrows, M. (1979). Synaptic potentials affect the release of transmitter from locust non-spiking interneurons. *Science* **204**, 81–3.

Burrows, M. (1980). The control of sets of motoneurones by local interneurones in the locust. *J Physiol* **298**, 213–33.

Burrows, M. (1985). The processing of mechanosensory information by spiking local interneurons in the locust. *J Neurophysiol* **54**, 463–78.

Burrows, M. (1989). Processing of mechanosensory signals in local reflex pathways of the locust. *J Exp Biol* **146**, 209–27.

Burrows, M. (1992a). Reliability and effectiveness of transmission from exteroceptive sensory neurones to spiking local interneurones in the locust. *J Neurosci* **12**, 1477–89.

Burrows, M. (1992b). Local circuits for the control of leg movements in an insect. *Trends Neurosci* **15**, 226–32.

Burrows, M. (1996). *The Neurobiology of an Insect Brain.* Oxford: Oxford University Press.

Burrows, M. & Newland, P.L. (1993). Correlation between the receptive fields of interneurons, their dendritic morphology, and the central projections of mechanosensory neurons. *J Comp Neurol* **329**, 412–26.

Burrows, M. & Pflüger, H.-J. (1995). Activity of locust neuromodulatory neurons is coupled to specific motor patterns. *J Neurophysiol* **74**, 47–57.

Burrows, M. & Siegler, M.V.S. (1978). Graded synaptic transmission between local interneurones and motor neurones in the metathoracic ganglion of the locust. *J Physiol* **285**, 231–55.

Burrows, M. & Siegler, M.V.S. (1985). The organization of receptive fields of spiking local interneurons in the locust with inputs from hair afferents. *J Neurophysiol* **53**, 1147–57.

Byrne, J.H. (1981). Comparative aspects of neural circuits for inking behavior and gill withdrawal in *Aplysia californica*. *J Neurophysiol* **45**, 98–106.

Byrne, J.H. & Kandel, E.R. (1996). Presynaptic facilitation revisited: state and time dependence. *J Neurosci* **16**, 425–35.

Camhi, J.M. & Tom, W. (1978). The escape system of the cockroach *Periplaneta americana*: I. The turning response to wind puffs. *J Comp Physiol* **128**, 193–201.

Canfield, J.G. & Rose, G.J. (1993). Activation of Mauthner neurons during prey capture. *J Comp Physiol A* **172**, 611–18.

Carew, T.J. & Kandel, E.R. (1977). Inking in *Aplysia californica*. I. Neural circuit of an all-or-nothing behavioral response. *J Neurophysiol* **40**, 692–707.

Carr, C.E. & Konishi, M. (1990). A circuit for detection of interaural time differences in the brain stem of the barn owl. *J Neurosci* **10**, 3227–46.

Catchpole, D.K. & Slater, P.J.B. (1995). *Bird Song: Biological Themes and Variations.* Cambridge: Cambridge University Press.

Comer, C.M. & Dowd, J.P. (1993). Multisensory processing for movement: antennal and cercal mediation of escape turning in the cockroach. In *Biological Neural*

Networks in Invertebrate Neuroethology and Robotics, ed. R.D. Beer, R.E. Ritzmann, & T. McKenna, pp. 89–112. Boston: Academic Press.

Cooke, I.R.C. & Macmillan, D.L. (1985). Further studies of crayfish escape behaviour. I. The role of the appendages and the stereotyped nature of non-giant escape swimming. *J Exp Biol* 118, 351–65.

Copenhaver, P.F. & Truman, J.W. (1982). The role of eclosion hormone in the larval ecdyses of *Manduca sexta*. *J Insect Physiol* 28, 695–701.

Cynx, J. (1990). Experimental determination of a unit of song production in the zebra finch (*Taeniopygia guttata*). *J Comp Psychol* 104, 3–10.

Dagan, D. & Camhi, J.M. (1979). Responses to wind recorded from the cercal nerve of the cockroach *Periplaneta americana*: II. Directional selectivity of the sensory nerves innervating single columns of filiform hairs. *J Comp Physiol A* 133, 103–10.

Diamond, J. (1968). The activation and distribution of gaba and L-glutamate receptors on goldfish Mauthner neurons: an analysis of dendritic remote inhibition. *J Physiol* 194, 669–723.

Doupe, A.J. (1997). Song- and order-selective neurons in the songbird anterior forebrain and their emergence during vocal development. *J Neurosci* 17, 1147–67.

Dowling, J.E. (1970). Organization of the vertebrate retina. *Invest Ophthalmol* 9, 655–80.

Dowling, J.E. (1987). *The Retina: an Approachable Part of the Brain*. Cambridge, MA: Belknap Harvard University Press.

Dvorak, D.R., Bishop, L.G. & Eckert, H.E. (1975). On the identification of movement detectors in the fly optic lobe. *J Comp Physiol* 100, 5–23.

Dyer, F.C. (1996). Spatial memory and navigation by honeybees on the scale of the foraging range. *J Exp Biol* 199, 147–54.

Eaton, R.C., DiDomenico, R. & Nissanov, J. (1991). Role of the Mauthner cell in sensorimotor integration by the brain stem escape network. *Brain Behav Evol* 37, 272–85.

Eaton, R.C. & Emberley, D.S. (1991). How stimulus direction determines the angle of the Mauthner initiated response in teleost fish. *J Exp Biol* 161, 469–87.

Eaton, R.C., Lavender, W.A. & Wieland, C.M. (1981). Identification of Mauthner-initiated response patterns in goldfish: evidence from simultaneous cinematography and electrophysiology. *J Comp Physiol* 144, 521–31.

Egelhaaf, M. (1985). On the neuronal basis of figure-ground discrimination by relative motion in the visual system of the fly. I. Behavioural constraints imposed by the neuronal network and the role of the optomotor system. *Biol Cybern* 52, 123–40.

Egelhaaf, M. & Borst, A. (1993). Motion computation and visual orientation in flies. *Comp Biochem Physiol* 104A, 659–73.

Erber, J., Masuhr, T. & Menzel, R. (1980). Localization of short-term memory in the bee of the brain, *Apis mellifera*. *Physiol Entomol* 5, 343–58.

Evans, P.D. & Siegler, M.V.S. (1982). Octopamine mediated relaxation of maintained and catch tension in locust skeletal muscle. *J Physiol* **324**, 93–112.

Ewert, J.-P. (1980). *Neuroetholgy*. Berlin: Springer-Verlag.

Ewert, J.-P. (1983). Neuroethological analysis of the innate releasing mechanism for prey-catching behaviour in toads. In *Advances in Vertebrate Neuroethology*, ed. J.-P. Ewert, R.R. Capranica & D.J. Ingle, pp. 413.75. New York: Plenum Press.

Ewert, J.-P. (1985). Concepts in vertebrate neuroethology. *Anim Behav* **33**, 1–29.

Ewert, J.-P. (1987). Neuroethology of releasing mechanisms: prey-catching in toads. *Behav Brain Sci* **10**, 337–403.

Falk, C.X., Wu, J.-Y., Cohen, L.B. & Tang, A.C. (1993). Nonuniform expression of habituation in the activity of distinct classes of neurons in the *Aplysia* abdominal ganglion. *J Neurosci* **13**, 4072–81.

Franceschini, N., Riehle, A. & Le Nestour, A. (1989). Directionally selective motion detection by insect neurons. In *Facets of Vision*, ed. D.G. Stavenga & R.C. Hardie, pp. 360–390. Berlin: Springer-Verlag.

Fraser, K. & Heitler, W.J. (1991). Photoinactivation of the crayfish segmental giant neuron reveals a direct giant-fiber to fast-flexor connection with a chemical component. *J Neurosci* **11**, 59–71.

Frost, W.N. & Katz, P.S. (1996). Single neuron control over a complex motor pattern. *Proc Natl Acad Sci USA* **93**, 422–6.

Furshpan, E.J. & Potter, D.D. (1959). Transmission at the giant motor synapses of the crayfish. *J Physiol* **145**, 289–325.

Gammie, S.C.G. & Truman, J.W. (1997). Neuropeptide hierarchies and the activation of sequential motor behaviors in the hawkmoth, *Manduca sexta*. *J Neurosci* **17**, 4389–97.

Glanzman, D.L. (1995). The cellular basis of classical conditioning in *Aplysia californica* – it's less simple than you think. *Trends Neurosci* **18**, 30–5.

Gnatzy, W. & Schmidt, K. (1971). Die Feinstruktur der Sinneshaare auf den Cerci von *Gryllus bimaculatus* Deg. (Saltatoria, Gryllidae). I. Faden und Keulenhaare. *Z Zellforsch* **122**, 190–209.

Grillner, S., Deliagina, T., Ekeberg, O. *et al.* (1995). Neural networks that co-ordinate locomotion and body orientation in lamprey. *Trends Neurosci* **18**, 270–9.

Grinnell, A.D. & Hagiwara, S. (1972). Studies of auditory neurophysiology in non-echolocating bats, and adaptations for echolocation in one genus, *Rousettus*. *Z vergl Physiol* **76**, 82–96.

Grinnell, A.D. & Schnitzler, H.-U. (1977). Directional sensitivity of echolocation in the horseshoe bat , *Rhinolophus ferrumequinum*. II. Behavioural directionality of hearing. *J Comp Physiol* **116**, 63–76.

Habersetzer, J. & Vogler, B. (1983). Discrimination of surface-structured targets by the echolocating bat *Myotis myotis* during flight. *J Comp Physiol* **152**, 275–82.

Hammer, M. (1993). An identified neuron mediates the unconditioned stimulus in associative olfactory learning in honeybees. *Nature* **366**, 59–63.

Hammer, M. & Menzel, R. (1995). Learning and memory in the honeybee. *J Neurosci* **15**, 1617–30.

Hardie, R.C. (1986). The photoreceptor array of the dipteran retina. *Trends Neurosci* **9**, 419–23.

Harrow, I.D., Hue, B., Pelhate, M. & Sattelle, D.B. (1980). Cockroach giant interneurones stained by cobalt-backfilling of dissected axons. *J Exp Biol* **84**, 341–3.

Hausen, K. & Egelhaaf, M. (1989). Neural mechanisms of visual course control in insects. In *Facets of Vision*, ed. D.G. Stavenga & R.C. Hardie, pp. 391–424. Berlin: Springer-Verlag.

Hawkins, R.D., Kandel, E.R. & Siegelbaum, S.A. (1993). Learning to modulate neurotransmitter release: themes and variations in synaptic plasticity. *Ann Rev Neurosci* **16**, 625–65.

Hedwig, B. & Pearson, K.G. (1984). Patterns of synaptic input to identified flight motoneurons in the locust. *J Comp Physiol A* **154**, 745–60.

Heitler, W.J. (1990). Electrical synapses. In *Neuronal Communications*, ed. W. Winlow, pp. 28–52. Manchester: Manchester University Press.

Heitler, W.J. & Fraser, K. (1993). Thoracic connections between crayfish giant fibres and motor giant neurones reverse abdominal patterns. *J Exp Biol* **181**, 329–33.

Hennig, M. (1990). Neuronal control of the forewings in two different behaviours in the cricket, *Teleogryllus commodus. J Comp Physiol A.* **167**, 617–27.

Hensler, K. (1992). Neuronal co-processing of course deviation and head movements in locusts. I. Descending deviation detectors. *J Comp Physiol A* **171**, 257–71.

Hensler, K. & Rowell, C.H.F. (1990). Control of optomotor responses by descending deviation detector neurones in intact flying locusts. *J Exp Biol* **149**, 191–205.

Herrmann, K. & Arnold, A.P. (1991). The development of afferent projections to the robust archistiatal nucleus in male zebra finches: a quantitative electron microscope study. *J Neurosci* **11**, 2063–74.

Horsman, U., Heinzel, H.-G. & Wendler, G. (1983). The phasic influence of self-generated air current modulations on the locust flight motor. *J Comp Physiol* **150**, 427–38.

Jarvis, E.D. & Nottebohm, F. (1997). Motor-driven gene expression. *Proc Natl Acad Sci USA* **94**, 4097–102.

Judd, S.P.D. & Collett, T.S. (1998). Multiple stored views and landmark guidance in ants. *Nature* **392**, 710–14.

Judge, S. J. & Rind, F.C. (1997). The locust DCMD, a movement-detecting neurone tightly tuned to collision trajectories. *J Exp Biol* **200**, 2209–16.

Kandel, E.R. (1976). *Cellular Basis of Behavior*. San Francisco: Freeman.

Katz, P.S., Getting, P.A. & Frost, W.N. (1994). Dynamic neuromodulation of synaptic strength intrinsic to a central pattern generator circuit. *Nature* **367**, 729–31.

Kimmel, C.B. & Eaton, R.C. (1976). Development of the Mauthner cell. In *Simpler Networks and Behavior*, ed. J.C. Fentress, pp. 186–202. Sunderland, MA: Sinauer Associates.

Kimmerle, B., Warzecha, A.-K. & Egelhaaf, M. (1997). Object detection in the fly during simulated translatory flight. *J Comp Physiol A* **181**, 247–55.

Kirchner, W.H. & Srinivasan, M.V. (1989). Freely flying honeybees use image motion to estimate object distance. *Naturwissenschaften* **76**, 281–2.

Kirn, J.R. & DeVoogd, T.J. (1989). Genesis and death of vocal control neurons during sexual differentiation in the zebra finch. *J Neurosci* **9**, 3176–87.

Knudsen, E.I. (1981). The hearing of the barn owl. *Sci Am* **245 (6)**, 83–91.

Knudsen, E.I. (1983). Early auditory experience aligns the auditory map of space in the optic tectum of the barn owl. *Science* **222**, 939–42.

Knudsen, E.I. (1998). Capacity for plasticity in the adult owl auditory system expanded by juvenile experience. *Science* **279**, 1531–3.

Knudsen, E.I., Blasdel, G.G. & Konishi, M. (1979). Sound localisation by the barn owl (*Tyto alba*) measured with the search coil technique. *J Comp Physiol* **133**, 1–11.

Knudsen, E.I. & Konishi, M. (1979). Mechanisms of sound localisation in the barn owl (*Tyto alba*). *J Comp Physiol* **133**, 13–21.

Kolton, L. & Camhi, J.M. (1995). Cartesian representation of stimulus direction: parallel processing by two sets of giant interneurons in the cockroach. *J Comp Physiol* **176**, 691–702.

Konishi, M. (1965a). Effects of deafening on song development in American robins and black-headed grosbeaks. *Z Tierpsychol* **22**, 584–99.

Konishi, M. (1965b). The role of auditory feedback in the control of vocalization in the white-crowned sparrow. *Z Tierpsychol* **22**, 770–83.

Konishi, M. (1992). The neural algorithm for sound localisation in the owl. *The Harvey Lectures* **86**, 47–64.

Konishi, M. (1993). Listening with two ears. *Sci Am* **268 (4)**, 34–41.

Koppl, C., Gleich, O. & Manley, G.A. (1993). An auditory fovea in the barn owl cochlea. *J Comp Physiol A* **171**, 695–704.

Krapp, H.G. & Hengstenberg, R. (1996). Estimation of self-motion by optic flow processing in single visual interneurons. *Nature* **384**, 463–6.

Krasne, F.B. (1969). Excitation and habituation of the crayfish escape reflex: the depolarising response in lateral giant fibres of the isolated abdomen. *J Exp Biol* **50**, 29–46.

Krasne, F.B. & Teshiba, T.M. (1995). Habituation of an invertebrate escape reflex due to modulation by higher centers rather than local events. *Proc Natl Acad Sci USA* **92**, 3362–6.

Krasne, F.B. & Wine, J.J. (1975). Extrinsic modulation of crayfish escape and behaviour. *J Exp Biol* **63**, 433–50.

Krasne, F.B. & Wine, J.J. (1977). Control of crayfish escape behavior. In *Identified Neurons and Behavior of Arthropods*, ed. G. Hoyle, pp. 275–92. New York: Plenum Press.

Kravitz, E. (1988). Hormonal control of behavior: amines and the biasing of neuronal output in the lobster. *Science* **241**, 1775–81.

Kupferman, I. & Weiss, K.R. (1978). The command neuron concept. *Brain Behav Sci* **1**, 3–39.

Kutsch, W. (1969). Neuromuskulare Aktivität bei verschiedenen Verhaltensweisen von drei Grillenarten. *Z Vergl Physiol* **63**, 335–78.

Kutsch, W., Schwarz, G., Fischer, H. & Kautz, H. (1993). Wireless transmission of muscle potentials during free flight of a locust. *J Exp Biol* **185**, 367–73.

Land, M.F. & Collett, T.S. (1974). Chasing behaviour of houseflies (*Fannia canicularis*). A description and analysis. *J Comp Physiol* **89**, 331–57.

Laughlin, S.B. (1981). Neural principles in the peripheral visual systems of invertebrates. In *Comparative Physiology and Evolution of Vision in Invertebrates: Invertebrate Visual Centers and Behavior I, Handbook of Sensory Physiology*, ed. H. Autrum, pp. 133–280. Berlin: Springer-Verlag.

Laughlin, S.B. (1994). Matching coding, circuits, cells and molecules to signals: general principles in retinal design of the fly's eye. *Prog Retin and Eye Res* **13**, 165–96.

Laughlin, S.B. & Hardie, R.C. (1978). Common strategies for light adaptation in the peripheral visual system of fly and dragonfly. *J Comp Physiol* **128**, 319–40.

Laughlin, S.B., Howard, J. & Blakeslee, B. (1987). Synaptic limitation to contrast coding in the retina of the blowfly *Calliphora*. *Proc R Soc B* **231**, 437–67.

Laughlin, S.B. & Weckström, M. (1993). Fast and slow photoreceptors – a comparative study of the functional diversity of coding and conductances in the Diptera. *J Comp Physiol A* **172**, 593–609.

Levi, R. & Camhi, J.M. (1996). Producing directed behaviour: muscle activity patterns of the cockroach escape responses. *J Exp Biol* **199**, 563–8.

Levine, R. B., Morton, D. B. & Restifo, L.L. (1995). Remodelling of the insect nervous system. *Curr Opin Neurobiol* **5**, 28–35.

Levine, R.B. & Truman, J.W. (1982). Metamorphosis of the insect nervous system: changes in morphology and synaptic interactions of identified neurones. *Nature* **299**, 250–2.

Levine, R.B. & Weeks, J.C. (1990). Hormonally mediated changes in simple reflex circuits during metamorphosis in *Manduca sexta*. *J Neurobiol* **21**, 1022–36.

Liebenthal, E., Uhlman, O. & Camhi, J.M. (1994). Critical parameters of the spike trains in a cell assembly: coding of turn direction by the giant interneurons of the cockroach. *J Comp Physiol A* **174**, 281–96.

Lillywhite, P.G. (1977). Single photon signals and transduction in an insect eye. *J Comp Physiol* **122**, 189–200.

Lindauer, M. (1967). Recent advances in bee communication and orientation. *Ann Rev Entomol* **12**, 439–70.

Link, A., Marimuthu, G. & Neuweiler, G. (1986). Movement as a specific stimulus for prey catching behaviour in rhinolophid and hipposiderid bats. *J Comp Physiol A* **159**, 403–13.

Lockery, S.R. & Kristan, W.B. (1990a). Distributed processing of sensory information in the leech. I . Input–output relations of the local bending reflex. *J Neurosci* **10**, 1811–15.

Lockery, S.R. & Kristan, W.B. (1990b). Distributed processing of sensory information in the leech. II. Identification of interneurons contributing to the local bending reflex. *J Neurosci* **10**, 1816–29.

Lockery, S.R., Wittenberg, G., Kristan, W.B. & Cottrell, G.W. (1989). Function of identified interneurons in the leech elucidated using neural networks trained by back-propagation. *Nature* **340**, 468–71.

Macdonald, D., ed. (1984). *The Encyclopedia of Mammals*. London: Allen & Unwin.

Manley, G.A., Koppl, C. & Konishi, M. (1988). A neural map of interaural intensity differences in the brain stem of the barn owl. *J Neurosci* **8**, 2665–76.

Mann, D.W. & Chapman, K.M. (1975). Component mechanisms of sensitivity and adaptation in an insect mechanoreceptor. *Brain Res* **97**, 331–6.

Margoliash, D. (1997). Functional organization of forebrain pathways for song production and perception. *J Neurobiol* **33**, 671–93.

Marler, P. & Tamura, M. (1964). Song 'dialects' in three species of white-crowned sparrow. *Science* **146**, 1483–6.

Masino, T. & Knudsen, E.I. (1990). Horizontal and vertical components of head movement are controlled by distinct neural circuits in the barn owl. *Nature* **345**, 434–7.

Mauelshagen, J. (1993). Neural correlates of olfactory learning paradigms in an identified neuron in the honeybee brain. *J Neurophysiol* **69**, 609–25.

McCasland, J.S. & Konishi, M. (1981). Interaction between auditory and motor activities in an avian song control nucleus. *Proc Natl Acad Sci USA* **78**, 7815–9.

Menzel, R. (1990). Learning, memory and 'cognition' in honeybees. In *Neurobiology of Comparative Cognition*, ed. R.P. Kesner & D.S. Olten, pp. 237–92. Hillsdale, N.J.: Erlbaum.

Meyrand, P., Simmers, A.J. & Moulins, M. (1991). Construction of a pattern generating circuit with neurons of different networks. *Nature* **351**, 60–3.

Meyrand, P., Simmers, A.J. & Moulins, M. (1994). Dynamic construction of a neural network from multiple pattern generators in the lobster stomatogastric nervous system. *J Neurosci* **14**, 630–44.

Miles, C.I. & Weeks, J.C. (1991). Developmental attenuation of the pre-ecdysis motor pattern in the tobacco hornworm, *Manduca sexta*. *J Comp Physiol A* **140**, 179–90.

Miller, J.P. & Selverston, A.I. (1982). Mechanisms underlying pattern generation in lobster stomatogastric ganglion as determined by selective inactivation of identified neurons. IV. Network properties of pyloric system. *J Neurophysiol* **48**, 1416–32.

Mizunami, M., Weibrecht, J.M. & Strausfeld, N.J. (1993). A new role for the insect mushroom bodies: place memory and motor control. In *Biological Neural Networks in Invertebrate Neuroethology and Robotics*, ed. R.D. Beer, R.E. Ritzmann & T. McKenna, pp. 199–226. Boston: Academic Press

Möhl, B. (1985). The role of proprioception in locust flight control. II. Information relayed by forewing stretch receptors during flight. *J Comp Physiol A.* **156**, 103–16.

Möhl, B. (1988). Short-term learning during flight control in *Locusta migratoria*. *J Comp Physiol A* **163**, 803–12.

Möhl, B. (1993). The role of proprioception for motor learning in locust flight. *J Comp Physiol A* **172**, 325–32.

Möhl, B. & Bacon, J. (1983). The tritocerebral commissure giant (TCG) wind-sensitive interneurone in the locust. II. Directional sensitivity and role in flight stabilisation. *J Comp Physiol* **150**, 453–65.

Moiseff, A. & Konishi, M. (1981). Neuronal and behavioural sensitivity to binaural time differences in the owl. *J Neurosci* **1**, 40–8.

Nagayama, T. (1989). Morphology of a new population of local interneurones in the locust metathoracic ganglion. *J Comp Neurol* **283**, 189–211.

Neuweiler, G. (1983). Echolocation and adaptivity to ecological constraints. In *Neuroethology and Behavioural Physiology*, ed. F. Huber, & H. Markl, pp. 280–302. Berlin: Springer-Verlag

Neuweiler, G., Bruns, V. & Schuller, G. (1980). Ears adapted for the detection of motion, or how echolocating bats have exploited the capacities of the mammalian auditory system. *J Acoust Soc Am* **68**, 741–53.

Neuweiler, G., Metzner, W., Heilmann, U., Rubsamen, R., Eckrich, M. & Costa, H. H. (1987). Foraging behaviour and echolocation in the rufous horseshoe bat (*Rhinolophus rouxi*) of Sri Lanka. *Behav Ecol Sociobiol* **20**, 53–67.

Neuweiler, G., Singh, S. & Sripathi, K. (1984). Audiograms of a South Indian bat community. *J Comp Physiol* **154**, 133–42.

Newland, P.L. (1991). Morphology and somatotopic organisation of the central projections of afferents from tactile hairs on the hind leg of the locust. *J Comp Neurol* **311**, 1–16.

Nicol, D. & Meinertzhagen, I.A. (1982). An analysis of the number and composition of synaptic populations formed by photoreceptors of the fly. *J Comp Neurol* **207**, 29–44.

Norberg, R.A. (1970). Hunting technique of Tengmalm's owl, *Aegolius funereus* (L.). *Ornis Scand* **1**, 51–64.

Norberg, R.A. (1977). Occurrence and independent evolution of bilateral ear asymmetry in owls and implications on owl taxonomy. *Phil Trans R Soc Lond B* **280**, 375–408.

Nottebohm, F. (1989). From bird song to neurogenesis. *Sci Am* **214**, 1368–70.

Nottebohm, F., Stokes, T. & Leonard, C. (1976). Central control of song in the canary. *J Comp Neurol* **165**, 457–86.

Novicki, A. & Weeks, J.C. (1995). A single pair of interneurons controls motor neuron activity during pre-ecdysis compression behaviour in larval *Manduca sexta*. *J Comp Physiol A.* **176**, 45–54.

O'Neill, W.E. & Suga, N. (1982). Encoding of target range and its representation in the auditory cortex of the moustached bat. *J Neurosci* **2**, 17–31.

O'Shea, M. & Rowell, C.H.F. (1976). The neuronal basis of a sensory analyser, the acridid movement detector system. II. Response decrement, convergence and the nature of the excitatory afferents to the fan-like dendrites of the LGMD. *J Exp Biol* **65**, 289–308.

Olson, G.C. & Krasne, F.B. (1981). The crayfish lateral giants are command neurons for escape behavior. *Brain Res.* **214**, 89–100.

Orchard, I., Ramirez, J.-M. & Lange, A.B. (1993). A multifunctional role for octopamine in locust flight. *Ann Rev Entomol* **38**, 227–49.

Payne, R.S. (1971). Acoustic location of prey by barn owls (*Tyto alba*). *J Exp Biol* **54**, 535–73.

Pearson, K.G. & Ramirez, J.-M. (1990). Influence of input from the forewing stretch receptors on motoneurones in flying locusts. *J Exp Biol* **151**, 317–40.

Pearson, K.G., Reye, D.N., Parsons, D.W. & Bicker, G. (1985). Flight-initiating interneurons in the locust. *J Neurophysiol* **53**, 910–23.

Pearson, K.G. & Wolf, H. (1987). Comparison of motor patterns in the intact and deafferented flight motor system of the locust. *J Comp Physiol A* **160**, 259–68.

Pearson, K.G. & Wolf, H. (1988). Connections of hindwing tegulae with flight neurones in the locust, *Locusta migratoria*. *J Exp Biol* **135**, 381–409.

Plummer, M. & Camhi, J.M. (1981). Discrimination of sensory signals from noise in the escape system of the cockroach: the role of wind acceleration. *J Comp Physiol* **142**, 347–57.

Pollack, G.D. (1980). Organizational and encoding features of single neurons in the inferior colliculus of bats. In *Animal Sonar Systems*, ed. R.G. Busnel & J.F. Fish, pp. 549–587. New York: Plenum Press.

Pollak, G.D., Marsh, D., Bodenhamer, R. & Souther, A. (1977). Characteristics of phasic-on neurons in the inferior colliculus of unanaesthetised bats with observations relating to mechanisms of echo ranging. *J Neurophysiol* **40**, 926–42.

Pollak, G.D. & Schuller, G. (1981). Tonotopic organization and encoding features of single units in inferior colliculus of horseshoe bats: functional implications for prey identification. *J Neurophysiol* **45**, 208–26.

Pringle, J.W.S. (1975). *Insect Flight*. Oxford Biology Readers No. 52. Oxford: Oxford University Press.

Prugh, J.I., Kimmel, C.B. & Metcalfe, W.K. (1982). Noninvasive recording of the Mauthner neurone action potential in larval zebra fish. *J Exp Biol* **101**, 83–92.

Ramirez, J.-M. & Orchard, I. (1990). Octopaminergic modulation of the fore-wing stretch receptor in the locust *Locusta migratoria*. *J Exp Biol* **149**, 255–79.

Ramirez, J.-M. & Pearson, K.G. (1991). Octopaminergic modulation of interneurons in the flight system of the locust. *J Neurophysiol* **66**, 1522–37.

Ramon y Cajal, S. (1911). *Histologie du Système Nerveux de l'Homme et des Vertébrés*. Madrid: Instituto Ramon y Cajal.

Reichert, H. & Wine, J.J. (1983). Coordination of lateral giant and non-giant escape systems in crayfish escape behaviour. *J Comp Physiol* **153**, 3–15.

Reichert, H., Wine, J.J. & Hagiwara, G. (1981). Crayfish escape behavior: neurobehavioral analysis of phasic extension reveals dual systems for motor control. *J Comp Physiol* **142**, 281–94.

Rind, F.C. (1984). A chemical synapse between two motion detecting neurones in the locust brain. *J Exp Biol* **110**, 143–67.

Rind, F.C. (1996). Intracellular characterization of neurons in the locust brain signalling impending collision. *J Neurophysiol* **75**, 986–95.

Rind, F.C. & Bramwell, D.I. (1996). A neural network based on the input organisation of an identified neuron signalling impending collision. *J Neurophysiol* **75**, 967–85.

Rind, F.C. & Simmons, P.J. (1992). Orthopteran DCMD neuron: a reevaluation of responses to moving objects. I. Selective responses to approaching objects. *J Neurophysiol* **68**, 1654–66.

Rind, F.C. & Simmons, P.J. (1998). A local circuit for the computation of object approach by an identified visual neuron in the locust. *J Comp Neurol* **395**, 405–15.

Riquimaroux, H., Gaioni, S. J. & Suga, N. (1991). Cortical computational maps control auditory perception. *Science* **251**, 565–8.

Ritzmann, R.E. (1993). The neural organization of cockroach escape and its role in context-dependent orientation. In *Biological Neuronal Networks in Invertebrate Neuroethology and Robotics.*, ed. R.D. Beer, R.E. Ritzmann & T. McKenna, pp. 113–37. New York: Academic Press.

Robert, D. (1989). The auditory behaviour of flying locusts. *J Exp Biol* **147**, 279–301.

Roberts, A. (1990). How does a nervous system produce behaviour? A case study in neurobiology. *Sci Prog* **74**, 31–51.

Roberts, A. & Tunstall, M.J. (1990). Mutual re-excitation with post-inhibitory rebound: a simulation study of the mechanisms of locomotor rhythm generation in the spinal cord of *Xenopus* embryos. *Eur J Neurosci* **2**, 11–23.

Robertson, R.M. & Pearson, K.G. (1982). A preparation for the intracellular analysis of neuronal activity during flight in the locust. *J Comp Physiol* **146**, 311–20.

Robertson, R.M. & Pearson, K.G. (1983). Interneurons in the flight system of the locust: distribution, properties and resetting properties. *J Comp Neurol* **215**, 33–50.

Robertson, R.M. & Pearson, K.G. (1985). Neural circuits in the flight system of the locust. *J Neurophysiol* **53**, 110–28.

Rossel, S. (1979). Regional differences in photoreceptor performance in the eye of the praying mantis. *J Comp Physiol* **131**, 95–112.

Rossel, S. (1980). Foveal fixation and visual tracking in the praying mantis. *J Comp Physiol* **139**, 307–31.

Sales, G. & Pye, D. (1974). *Ultrasonic Communication by Animals*. London: Chapman and Hall.

Schramek, J.E. (1970). Crayfish swimming: alternating motor output and giant fiber activity. *Science* **169**, 698–700.

Schuller, G. (1984). Natural ultrasonic echoes from wing beating insects are encoded by collicular neurons in the CF-FM bat, *Rhinolophus ferrumequinum*. *J Comp Physiol* **154**, 121–8.

Selverston, A.I. & Miller, J.P. (1980). Mechanisms underlying pattern generation in lobster stomatogastric ganglion as determined by selective inactivation of identified neurons. I. Pyloric system. *J Neurophysiol* **44**, 1102–21.

Shepherd, G.M. (1983). *Neurobiology*. New York: Oxford University Press.

Siegler, M.V.S. (1981). Postural changes alter synaptic interactions between non-spiking interneurons and motor neurons in the locust. *J Neurophysiol* **46**, 310–23.

Siegler, M.V.S. & Burrows, M. (1984). The morphology of two groups of spiking local interneurons in the metathoracic ganglion of the locust. *J Neurosci* **6**, 507–13.

Siegler, M.V.S. & Burrows, M. (1986). Receptive fields of motor neurones underlying local tactile reflexes in the locust. *J Neurosci* **6**, 507–13.

Sillar, K., Wedderburn, J.F.S. & Simmers, A.J. (1991). The postembryonic development of locomotor rhythmicity in *Xenopus laevis* tadpoles. *Proc R Soc B* **246**, 147–53.

Simmons, P.J. & Rind, F.C. (1992). Orthopteran DCMD neuron: a reevaluation of responses to moving objects. II. Critical cues for detecting approaching objects. *J Neurophysiol* **68**, 1667–82.

Snodgrass, R.E. (1935). *Principles of Insect Morphology*. New York: McGraw-Hill.

Snyder, A.W., Stavenga, D.G. & Laughlin, S.B. (1977). Spatial information capacity of compound eyes. *J Comp Physiol* **116**, 183–207.

Solis, M.M. & Doupe, A.J. (1997). Anterior forebrain neurons develop selectively by an intermediate stage of birdsong learning. *J Neurosci* **17**, 6447–62.

Spinola, S.M. & Chapman, K.M. (1975). Proprioceptive indentation of the campaniform sensilla of cockroach legs. *J Comp Physiol* **96**, 257–72.

Srinivasan, M.V. (1992). How bees exploit optic flow: behavioural experiments and neural models. *Phil Trans R. Soc B* **337**, 253–9.

Srinivasan, M.V., Laughlin, S.B. & Dubs, A. (1982). Predictive coding: a fresh view of inhibition in the retina. *Proc R Soc B* **216**, 427–59.

Sterling, P. (1998). Retina. In *The Synaptic Organization of the Brain*, ed. G.M. Shepherd, pp. 205–53. Oxford: Oxford University Press.

Streichert, L.C. & Weeks, J.C. (1995). Decreased monosynaptic input to an identified motoneuron is associated with steroid-mediated dendritic regression during metamorphosis in *Manduca sexta*. *J Neurosci* **15**, 1484–95.

Suga, N., Neuweiler, G. & Moller, J. (1976). Peripheral auditory tuning for fine frequency analysis by the CF-FM bat, *Rhinolophus ferrumequinum*. IV. Properties of peripheral auditory neurons. *J Comp Physiol* **106**, 111–25.

Sullivan, W.E. (1982). Neural representation of target distance in auditory cortex of the echolocating bat, *Myotis lucifugus*. *J Neurophysiol* **48**, 1011–32.

Suthers, R.A. (1990). Contributions to birdsong from the left and right sides of the intact syrinx. *Nature* **347**, 473–7.

Syed, N.I., Bulloch, A.G.M. & Lukowiak, K. (1990). In vitro reconstruction of the respiratory central pattern generator of the mollusk *Lymnaea*. *Science* **250**, 282–5.

Takahashi, T., Moiseff, A. & Konishi, M. (1984). Time and intensity cues are processed independently in the auditory system of the owl. *J Neurosci* **4**, 1781–6.

Thorpe, W.H. (1958). The learning of song patterns by birds, with especial references to the song of the chaffinch, *Fringilla coelebs*. *Ibis* **100**, 535–70.

Tinbergen, N. (1951). *The Study of Instinct*. Oxford: Oxford University Press.

Tinbergen, N. (1963). On aims and methods of ethology. *Z Tierpsychol* **20**, 410–33.

Truman, J.W. & Reiss, S.E. (1988). Hormonal regulation of the shape of identified motoneurons in the moth *Manduca sexta*. *J Neurosci* **8**, 765–75.

Vater, M., Feng, A.S. & Betz, M. (1985). An HRP-study of the frequency–place map of the horseshoe bat cochlea: morphological correlates of the sharp tuning to a narrow frequency band. *J Comp Physiol A* **157**, 671–86.

Vates, G.E. & Nottebohm, F. (1995). Feedback circuitry within a song-learning pathway. *Proc Natl Acad Sci USA* **92**, 5139–43.

Vicario, D.S. (1991). Organization of the zebra finch song control system: II. Functional organization of the output from the nucleus robustus archistiriatalis. *J Comp Neurol* **309**, 486–94.

Vu, E.T., Mazurek, M.E. & Kuo, Y.-C. (1994). Identification of a forebrain motor programming network for the learned song of zebra finches. *J Neurosci* **14**, 6924–34.

Warzecha, A.-K., Egelhaaf, M. & Borst, A. (1993). Neural circuit tuning fly visual neurons to motion of small objects. I. Dissection of the circuit by pharmacological and photoinactivation techniques. *J Neurophysiol* **69**, 329–39.

Watkins, B.L., Burrows, M. & Siegler, M.V.S. (1985). The structure of locust nonspiking interneurones in relation to the anatomy of their segmental ganglion. *J Comp Neurol* **240**, 233–55.

Watson, A.H.D. (1984). The dorsal unpaired median neurons of the locust metathoracic ganglion: neuronal structure and diversity, and synapse distribution. *J Neurocytol* **13**, 303–27.

Watson, A.H.D. & Burrows, M. (1985). The distribution of synapses on the two fields of neurites of spiking local interneurones in the locust. *J Comp Neurol* **240**, 219–32.

Watson, A.H.D. & Burrows, M. (1988). The distribution and morphology of synapses on nonspiking local interneurones in the thoracic nervous system of the locust. *J Comp Neurol* **272**, 605–16.

Weeks, J.C. & Levine, R.B. (1995). Steroid effects on neurons subserving behaviour. *Curr Opin Neurobiol* **5**, 809–15.

Weeks, J.C. & Truman, J.W. (1984). Neural organization of peptide activated ecdysis behaviors during metamorphosis of *Manduca sexta*. *J Comp Physiol A* **155**, 407–22.

Wiersma, C.A.G. (1947). Giant nerve fiber system of the crayfish: a contribution to comparative physiology of the synapse. *J Neurophysiol* **10**, 23–38.

Wiersma, C.A. G. & Ikeda, K. (1964). Interneurons commanding swimmeret movements in the crayfish, *Procambarus clarkii* (Girard). *Comp Biochem Physiol* 12, 509–25.

Willows, A.O.D., Dorsett, D.A. & Hoyle, G. (1973). The neuronal basis of behavior in *Tritonia*. III. Neuronal mechanism of a fixed action pattern. *J Neurobiol* 4, 255–85.

Wilson, D.M. (1960). The central nervous control of flight in a locust. *J Exp Biol* 38, 471–90.

Wilson, M. (1978). The functional organization of locust ocelli. *J Comp Physiol* 124, 297–316.

Wilson, M., Garrard, P. & McGiness, S. (1978). The unit structure of the locust compound eye. *Cell Tiss Res* 195, 205–26.

Wine, J.J. (1977). Crayfish escape behaviour II. Command derived inhibition of abdominal extension. *J Comp Physiol* 121, 173–86.

Wine, J.J. (1984). The structural basis of an innate behaviour pattern. *J Exp Biol* 112, 283–319.

Wine, J.J. & Krasne, J.B. (1982). The cellular organization of crayfish escape behavior. In *The Biology of Crustacea*, ed. E.D. Bliss, pp. 241–292. New York: Academic Press.

Wine, J.J. & Mistick, D.C. (1977). Temporal organization of crayfish escape behavior: delayed recruitment of peripheral inhibition. *J Neurophysiol* 40, 904–25.

Wong, D., Maekawa, M. & Tanaka, H. (1992). The effect of pulse repetition rate on the delay sensitivity of neurons in the auditory-cortex of the fm bat, *Myotislucifugus*. *J Comp Physiol* 170, 393–402.

Wong, D. & Shannon, S. (1988). Functional zones in the auditory cortex of the echo-locating bat, *Myotis lucifugus*. *Brain Res* 453, 349–52.

Wu, J.-Y., Cohen, L.B. & Falk, C.X. (1994). Neuronal activity during different behaviors in *Aplysia*: a distributed organization? *Science* 263, 820–3.

Yu, A.C. & Margoliash, D. (1996). Temporal hierarchical control of singing in birds. *Science* 273, 1871–5.

Zečević, D., Wu, J.-Y., Cohen, L.B., London, J.A., Höpp, H.-P. & Falk, C.X. (1989). Hundreds of neurons in the *Aplysia* abdominal ganglion are active during the gill withdrawal reflex. *J Neurosci* 9, 3681–9.

Zill, S.N. & Moran, D.T. (1981). The exoskeleton and insect proprioception. I. Responses of campaniform sensilla to external and muscle-generated forces in the American cockroach, *Periplaneta americana*. *J Exp Biol* 91, 1–24.

Zitnan, D., Kingan, T.G., Hermesman, J.L. & Adams, M.E. (1996). Identification of ecdysis-triggering hormone from an epitracheal endocrine system. *Science* 271, 88–91.

Zottoli, S.J. (1977). Correlation of the startle reflex and Mauthner cell auditory responses in unrestrained goldfish. *J Exp Biol* 66, 243–54.

Zottoli, S.J. (1978). Comparative morphology of the Mauthner cell in fish and amphibians. In *Neurobiology of the Mauthner cell.*, ed. D.S. Faber & H. Korn, pp. 13–45. New York: Raven Press.

INDEX

Entries in **bold type** refer to pages where an important term, set in **bold** in the text, is defined or used for the first time.